Microtechnology and Thermal Problems in Electronics (MicroTherm'2015) and International Conference on Smart Engineering of New Materials (SENM'2015)

Journal of Physics: Conference Series Volume 709

Lodz, Poland
22 – 25 June 2015

Editors:

Ewa Raj
Katarzyna Znajdek

ISBN: 978-1-5108-2420-1
ISSN: 1742-6588

Printed from e-media with permission by:

Curran Associates, Inc.
57 Morehouse Lane
Red Hook, NY 12571

Some format issues inherent in the e-media version may also appear in this print version.

Printed by Curran Associates, Inc. (2016)

For permission requests, please contact the Institute of Physics
at the address below.

Institute of Physics
Dirac House, Temple Back
Bristol BS1 6BE UK

Phone: 44 1 17 929 7481
Fax: 44 1 17 920 0979

techtracking@iop.org

Additional copies of this publication are available from:

Curran Associates, Inc.
57 Morehouse Lane
Red Hook, NY 12571 USA
Phone: 845-758-0400
Fax: 845-758-2633
Email: curran@proceedings.com
Web: www.proceedings.com

TABLE OF CONTENTS

PREFACE

The issue covers selected papers presented during **the Joint Event** of two International Cyclic Conferences: **MicroTherm** - Microtechnology and Thermal Problems in Electronics and **SENM** - Smart Engineering of New Materials. The Duoconference was held in Lodz, Poland on 23-25 June 2015. It gathered almost 200 participants from 20 countries.

MicroTherm is an International Conference on Microtechnology and Thermal Problems in Electronics that is organised every two years, since 1996. The success of the first seminar devoted to thermal management aspects, and the successive conferences have led us to the eleventh edition. Since the first meeting, the scope of the Conference has expanded, following the progress of electronics. Now, it covers subjects connected with extreme temperature, electronics, sensors and measurement techniques, modelling, simulation, wide band-gap materials, packaging and reliability, renewable energy sources and photonics with special emphasis on microelectronic technologies.

The tradition of **SENM** dates back to 1985 when the First International Conference on Diamond Crystallization under Reduced Pressure was held in Jabłonna, Poland. Since then, the Conference is organised regularly every five years. From the early beginning the main aim of this event was the creation of an international forum of scientists for the presentation of the newest achievements and investigation results in vacuum and plasma techniques, synthesis, characterisation and application of diamond, its derivatives and other wide bandgap materials. Over time, the scope of the conference has expanded to include nanomaterials, biomaterials and new functional materials, in the range of production, design and optimisation of the synthesis and modification technologies.

Traditionally, the programme of **MicroTherm'2015** and **SENM'2015** is comprised of invited talks and regular sessions in the form of planar discussions and poster presentations, which also include a Students' Session. The latter gave an opportunity for students and young researchers to present their first achievements in the field of science.

Next Joint Event combining **MicroTherm** and **SENM** Conferences is going to be held in June 2017, traditionally in Lodz - beautiful, post-industrial city located in the centre of Poland. For further details, please see the Duoconference web site www.semtherm.eu.

Please, feel invited.

Ewa Raj and Katarzyna Znajdek
Editors

Journal of Physics: Conference Series **709** (2016) 011002 doi:10.1088/1742-6596/709/1/011002

Peer review statement

All papers published in this volume of *Journal of Physics: Conference Series* have been peer reviewed through processes administered by the proceedings Editors. Reviews were conducted by expert referees to the professional and scientific standards expected of a proceedings journal published by IOP Publishing.

Journal of Physics: Conference Series **709** (2016) 012001 doi:10.1088/1742-6596/709/1/012001

The influence of nanoadditives on the tribological properties of process fluids

T Bakalova, L Svobodová, K Bořůvková, P Louda, L Voleský

Institute for Nanomaterials, Advanced Technology and Innovation, Technical University of Liberec, Studentska 2, 461 17, Czech Republic

E-mail: totka.bakalova@tul.cz

Abstract. Tribology deals with interaction of surfaces in relative motion depending on their design, friction, wear and lubrication. The proper use of process fluids or lubricants can bring a significant reduction in friction and the amount of wear, thereby leading to a reduction in power consumption. During different technological operations contamination of used process fluids or lubricants occurs. Such contamination leads not only to a reduction of the lifetime of the lubricants but it can also change the functional properties and increase the health risks for operators. The quality of the process fluid is among other things influenced by bacterial attacks. The use of nanoadditives is one method for inhibiting the bacteria and improving the bioavailability and stability of the technological fluids. Nanolubricant is a new system composed of nanometer-sized particles dispersed in a base lubricant. The doping of lubricants with nanoparticles is one of the ways to solve problems with the removal of bacteria, whereby improving the biological, chemical and technological stability of process fluids. In the article, we monitor the effects of doping process fluids with nanoparticles of silica (SiO_2), titanium dioxide (TiO_2), silver nitrate ($AgNO_3$) and ascorbic acid ($C_6H_8O_6$) on the friction coefficient and on the wear of friction pairs of Si_3N_4 balls against steel 16MnCr5, EN 10084-94.

1. Introduction

Machining technology is still a rapidly developing sector in the engineering industry. Both machine tools and in particular cutting tools are experiencing rapid development and not the least in the area of process media [1-3]. An important component of the metal cutting process is fluid; its quality often decides the outcome of the machining process and the lifespan of a machine. The correct choice of process fluids leads to cost savings, enhanced machining performance and better product quality [4-7]. The various substances that are received in the process fluid are to be considered as contamination, as they adversely affect the composition of the process fluid, its performance and health characteristics. They can be divided into solid, fluid and biological contamination. In the case of newly produced process fluids strong emphasis is placed on ecology and health and safety in their use and removal. The order of Regulation (EC) No 1907/2006 of the European Parliament and of the Council (REACH - Registration, Evaluation and Authorisation of Chemicals) is applied and authorization of chemicals with ECHA (European Chemicals Agency) also takes place.

The lifetime (degradation) of process fluids in the first instance is affected by bacterial attack. The resistance of process fluids to bacterial attack determines the resulting lifetime of the fluid. The bacteria that attack process fluids are called Anaerobic, i.e. bacteria that multiply the most if no oxygen is present. Emulsifiers, sulfonates, fatty alcohols, corrosion inhibitors and other organic components serve primarily as food for these bacteria. An important component incorporated into process fluids are biocide additives. Their purpose is to protect the fluid against the biological attack by bacteria or fungi. Another problem that is solved is the protection of users against spores and endotoxins, which would otherwise be formed by these organisms. An increased proportion of these ingredients can lead to the improved lifetime of process media but on the other hand leads to allergic reactions in people who are exposed to the fluid [8, 9]. We are constantly looking for ways to solve these urgent problems. The main objective is to provide a safe environment for a person who in execution of his/her work is in constant contact with process fluids. Given the growing concerns about the environment and regulations on contamination and pollution there is an observable increase in the interest in renewable and biodegradable cutting fluids. In this respect, the significance of the term "green" machining, which includes both the type of process fluid and the methods to be applied during the machining process, is also increasing. [10-12]

The use of nanoadditives in the form of nanoparticles is highly efficient due to their high chemical and biological activity. Application of nanoparticles is one of the possible ways to solve problems with reduce the presence of bacteria and fungi, thus ensuring improved biological, chemical and technological stability of the process fluids. A suitable alternative for the development of antimicrobial solutions is to synthesize the biocide attributes (especially nanoparticles based on silver) of metal nanoparticles. Nanoparticles with dimensions smaller than 100 nm have a large surface area so they can interact easily and in a high rate with biological material. [10-14]

The intensity of the machining process is significantly increased and thus leads to an increase in the cutting speed. The most analyzed parameters are the surface roughness and hardness. In addition to the most analyzed parameters the tribological properties of the newly created functional surfaces are also increasingly analyzed. The tribology evaluates values that affect the mutual interaction of the surface environment and the body during their relative motion. The mechanism of friction can be defined as a resistance to the relative motion of two bodies during their contact. The friction force F depends on the friction coefficient μ and the loading force F_N. The description of friction and the introduction of this concept were first discussed by Coulomb (1799) and the basic relationship is known as Coulomb's law [15]. Sliding friction arises when two bodies come into contact with one another and leads to their movement relative to each other. In fact, the real contact between bodies occurs only in a limited number of small areas on the surfaces (Figure 1).

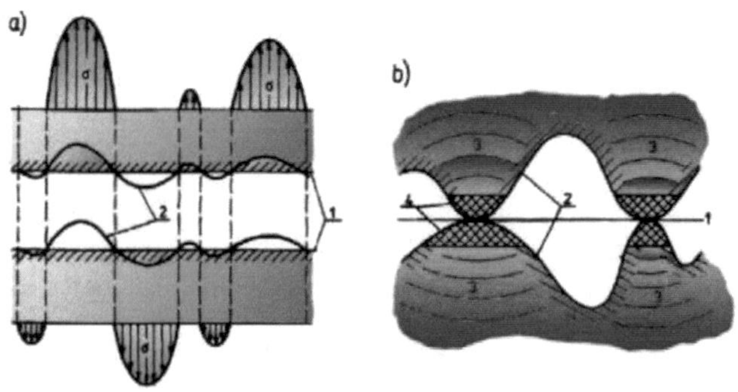

Figure 1. Schematic illustration of contact during sliding friction [16].

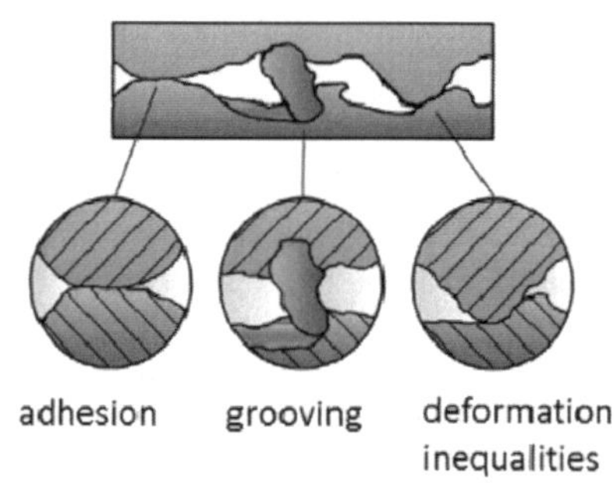

Figure 2. Components of sliding friction [17].

Figure 1 shows (a) body contact with an ideally flat surface and stress distribution (b) real contact, which includes: ideal surface - 1, real surface - 2, elastic deformation region - 3 and plastic deformation, region - 4. The size of the contact area increases with increasing load. On the tops a connection may occur, thus creating resistance to relative motion. In the specialized literature [16] a relation was derived for the coefficient of friction, according to the equation (1):

$$F_t = A_{mech} . \tau_s = \mu . F_N \quad (1)$$

where: F_t *[N]* – frictional force acting against the direction of movement;
A_{mech} *[mm²]* – actual contact area of the firm connection;
τ_s *[MPa]* – shear strength of the material;
F_N *[N]* – load force;
μ *[-]* – coefficient of friction.

In the case of sliding friction three basic components can be observed. The ratio of each component depends on the type of material pairs, the friction conditions (material properties of pairs of movements, surface roughness, presence of a lubricant etc.) and the time stage of the mutual movement.

The following mechanisms are applied when friction occurs (Figure 2):

- Friction due to adhesion between the surface roughness;
- friction due to striations caused by the presence of fine particles arising from the wear surface of friction due to deformation of the surface roughness;
- friction due to the deformation of the surface roughness. [17, 18]

2. Materials and Methods

2.1. Evaluation of tribological tests

The basis of tribological measurements is the *"ball-on-disc"* testing method. The measurement involves the injection of fixed attachments of a test piece ("ball") in the form of balls (Figure 3) of the chosen material with a defined force to drive (test sample). An essential part is the friction sensor. The coefficient of friction between the unit and the disc is determined during the test measurement. [7]

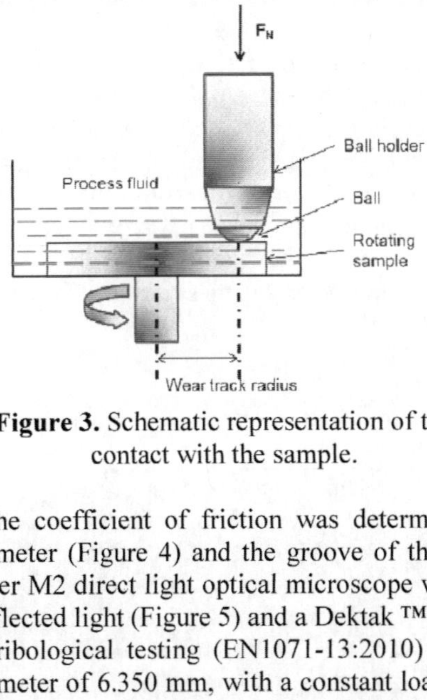

Figure 3. Schematic representation of the contact with the sample.

Figure 4. CETR UMI Multi-Specimen Test System.

The coefficient of friction was determined using a CETR UMI Multi-Specimen Test System tribometer (Figure 4) and the groove of the tribological test was evaluated by a (LM) Zeiss AXIO Imager M2 direct light optical microscope with a fully motorized stage configuration for observations in reflected light (Figure 5) and a Dektak ™ XT mechanical profilometer (Figure 6) from BRUKER.

Tribological testing (EN1071-13:2010) was conducted by using a ball made from Si_3N_4 with a diameter of 6.350 mm, with a constant load of 10 N at room temperature and humidity of 40 ± 2 %. The material of the disc was steel 16MnCr5 - EN 10084-94, EN 84-70 with a polished surface and a roughness of Ra 0.01 μm. The radius of the circle along the "ball" body was moving was 12 mm and the coefficient of friction was measured using a 5% solution of process fluids with a volume of 100 ml.

The use of a light-optical microscope allows the microscopic objects and structures to be observed and analyzed in reflected light at magnifications from 25x to 1000x. Preparation of the samples for analysis using the optical microscope in reflected light first involves grinding and then polishing. The contrast of the images is obtained based on the individual structural reflexivity, different phases or due to the surface topography. The intensity of reflected light is proportional to $cos\varphi$, where φ is the angle between the sample surface and the incident light beam. [19]

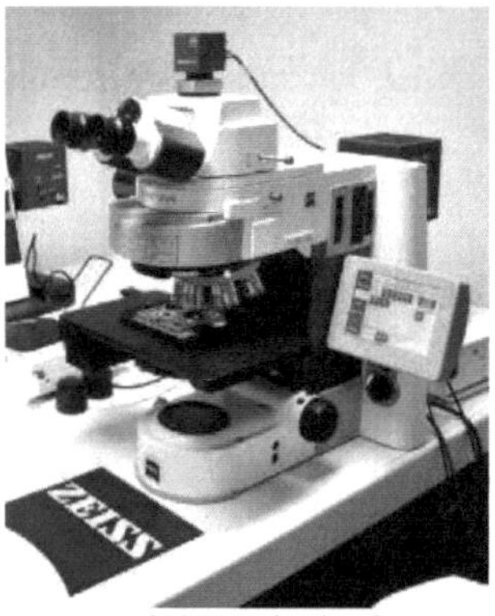

Figure 5. Light optical microscope.

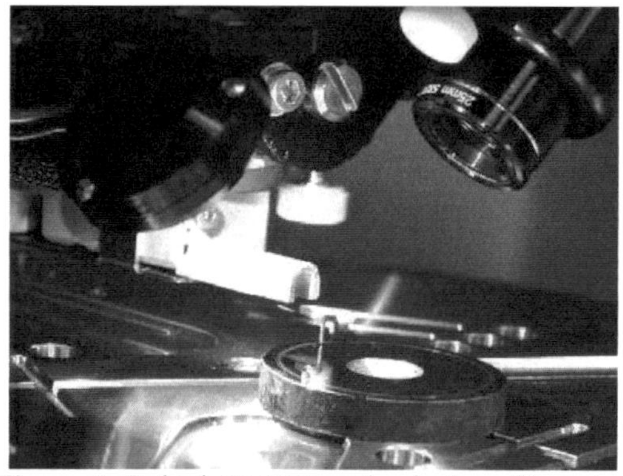

Figure 6. Dektak ™ XT mechanical profilometer.

A recording of images from a light optical microscope is obtained using a CCD camera. Software can then be used to archive analyse and modify the scanned data. The aim of the software data analysis is to process visual information, eliminate random errors and extract typical features characterizing the image [19, 20]. Optical microscopy allows structures to be displayed very fast using various contrast methods, e.g. phase contrast, polarization, fluorescence, Hoffman modulation contrast or Nomarski differential interference contrast [21].

The Dektak ™ XT mechanical profilometer comes into direct contact with the surface of the sample tip. A load must be selected according to the type of sample to prevent the mechanical damage (scratching) of samples if the sample is soft and the force is disproportionately large. During the measurement the sample is placed on a substrate and is in direct contact with the tip. The inequalities on the surface of the sample are registered by the tip which performs a linear movement, and the samples move relatively to the stationary tip. In the contact mode the mechanical profilometer scans the shape of the surface in contact with a diamond tip, which has the size of a few micrometres; thereby it is possible to study the surface morphology of the horizontal plane with the size from a hundred micrometres to tens of millimetres [7].

2.2. Use of natural fillers in the form of nanoparticles

When using cooling and lubricating process fluids contamination due to technological process occurs.

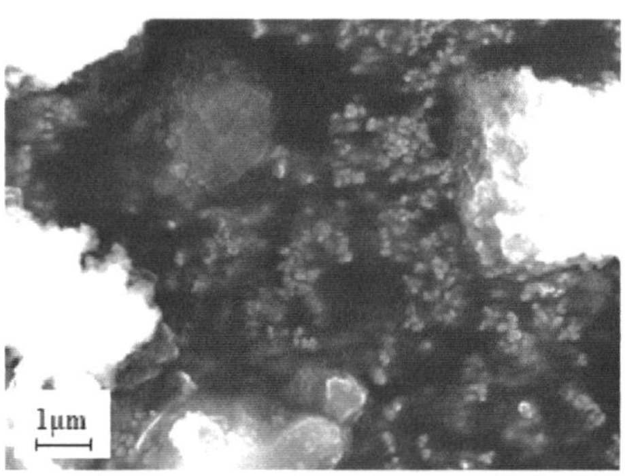

Figure 7. SEM images of Silicon dioxide (SiO_2) nanoparticles.

Figure 8. SEM images of Titanium dioxide (TiO_2) nanoparticles.

This has a negative impact on the reduction of their lifespan and also changes their functional properties. Bacterial attacks also have an impact on the quality of process fluids. The resistance of process fluids to bacterial attack determines the resulting lifespan of the fluid. The nanoparticles may be added to reduce the presence of bacteria and fungi, and if this occurs, prevent their spreading.

The aim of the modification of industrially supplied process fluids is to increase their biological stability even after long-term use. The experiment was performed at the Institute for Nanomaterials, Advanced Technology and Innovation, Technical University of Liberec (TUL), Czech Republic. Two standard supplied process fluids labelled PF 1 and PF 2 were used in the experiment.

Nanoparticles of silicon dioxide SiO_2 and titanium dioxide TiO_2 (Figure 7 and Figure 8) were added to the process fluids and subsequently tribological tests were carried out and the technological and biological stability of the processes were evaluated.

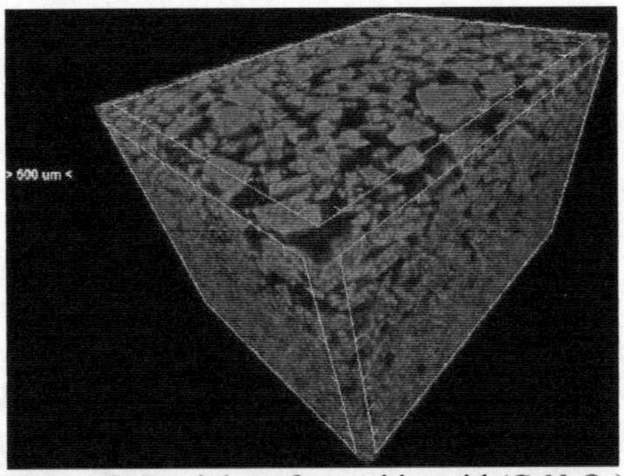

Figure 9. Particles of ascorbic acid ($C_6H_8O_6$) from a Sky Scan 1272 micro-tomograph device.

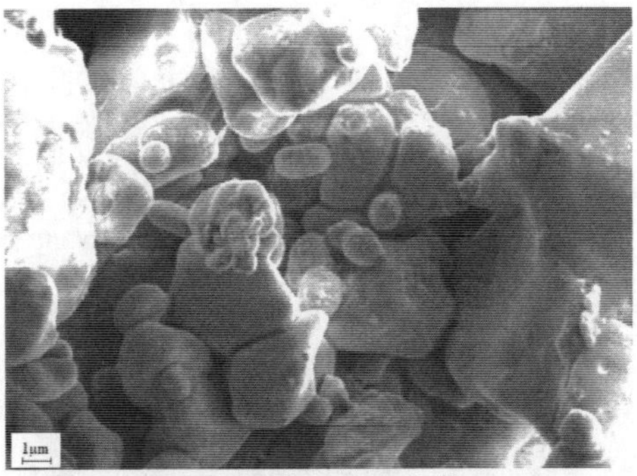

Figure 10. SEM images of silver nitrate ($AgNO_3$) nanoparticles.

Both process fluids were modified by the addition of silver nitrate ($AgNO_3$) and ascorbic acid ($C_6H_8O_6$). Ascorbic acid (Figure 9) was chosen for the purpose of improving antimicrobial and antioxidant activity and also to reduce the metallic silver nanoparticles from the silver nitrate solution (Figure 10). Variants of process fluids were prepared: with the addition of silver only and with the addition of silver and ascorbic acid. Reduction of silver nanoparticles from silver nitrate was carried out using UV radiation. UV irradiation was performed in a closed chamber using a high performance halogen metal vapour lamp with a combination of metal halides generating strong radiation in the UVB range (280 - 315 nm) and especially in the UVA range (315 - 380 nm). Reduction of silver nanoparticles resulted in a change in colour of the tested process fluids.

3. Experimental results

The effects of adding a nanoadditive to the process fluids on their tribological behaviour were evaluated by comparing the frictional properties and wear-resistant material for tribology. The friction coefficients without and after adding the nanoadditives to a 5% solution were evaluated as well as the tribological wear of the material after the experiment.

3.1. Evaluation of friction and wear using process fluid 1

Process fluid 1 is a water miscible high-performance cooling lubricant based on mineral oils. The examined PF 1 contains a high proportion of polar additives, during processing it has long-term stability and provides excellent corrosion protection.

The tribological results are shown in Figure 11 (left), where we can see a comparison of the friction coefficient of the pure 5% PF 1 solution and solutions with added nanoparticles (NPs) of Silicon dioxide (SiO_2), Titanium dioxide (TiO_2), Silver nitrate ($AgNO_3$) and Ascorbic acid ($C_6H_8O_6$), marked as Ag+AA NPs in the Fig. 11. Rating of ball wear was performed according to standard EN1071-

13:2010, friction pair were ball of Si_3N_4 and steel 16MnCr5-EN 10084-94, EN 84-70, results are shown in Figure 11 (right).

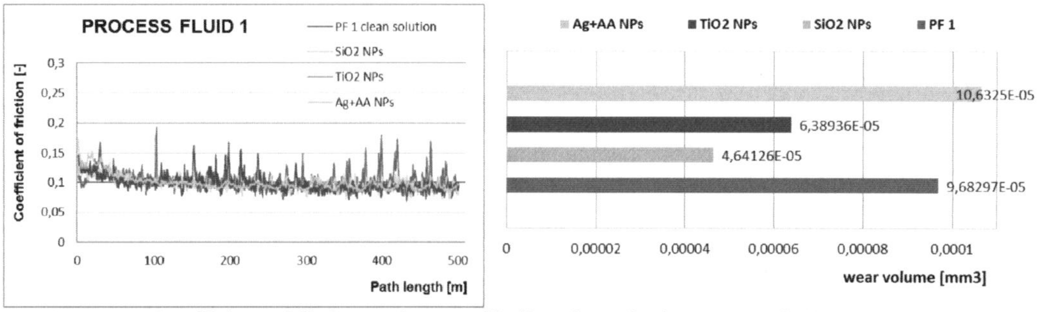

Figure 11. Coefficient of fiction and wear of balls (Si_3N_4) in the process fluid PF1 without/with nanoparticles.

With the help of the optical microscope several important parameters for wear can be evaluated, e.g. size of wear beads, which are shown in Figures 12 and 13 (top left of the figure).

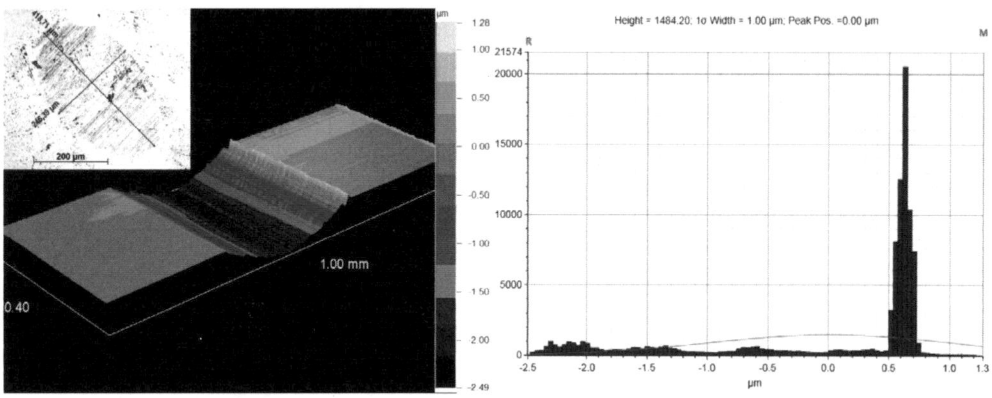

Figure 12. Mechanical profilometer: 3D image of profile on the disk after tribology - clean solution and optical microscopic images of wear of the ball.

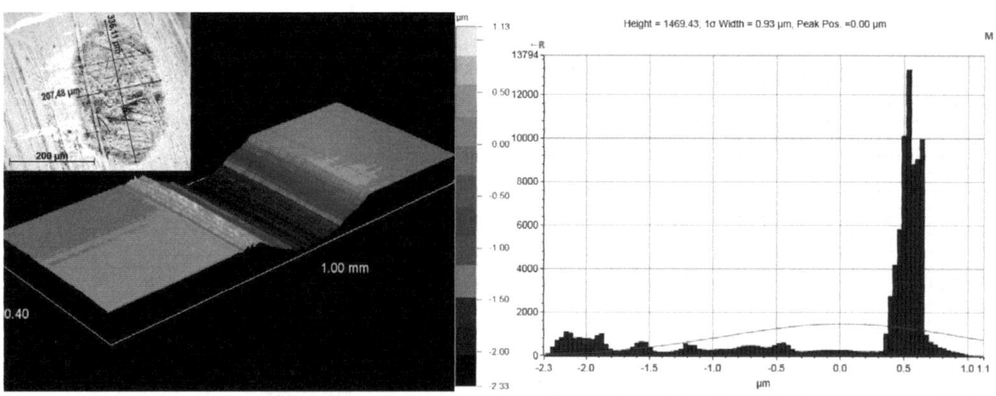

Figure 13. Mechanical profilometer: 3D image of profile on the disk after tribology - nanoadditive Silicon dioxide (SiO_2) and optical microscopic images of wear of the ball

Information on the volume of spalled material can be obtained using a Dektak XT ™ mechanical profilometer, Figures 12 and 13, which shows the wear when using the unmodified process fluid and process fluid with nanoparticles Silicon dioxide (SiO_2). The relevant software enables the measured values to be sorted using Gauss' law along the investigated profile and offers the possibility of sorting values in a histogram according to their frequency. The wear rate of the friction pair of two parts provides information on the friction properties of the media used under specific conditions (Figure 11 right - wear of ball Si_3N_4 without/with nanoparticles).

3.2. Evaluation of friction and wear using process fluid 2

Process fluid 2 is a semi-synthetic, water-miscible high-performance cooling lubricant based on mineral oils. The examined fluid PF 2 contains a high proportion of polar additives containing boron and bactericide; it has long-term stability and provides excellent corrosion protection.

The tribological results are shown in Figure 14 (left), where the coefficient of friction using a 5% solution of pure PF 1 is compared with solutions with the addition of nanoparticles (NPs) of silicon dioxide (SiO_2), titanium dioxide (TiO_2) Silver nitrate ($AgNO_3$) and ascorbic acid ($C_6H_8O_6$), marked as Ag+AA NPs in the Fig. 14. Rating of ball wear was performed according to standard EN1071-13:2010, friction pair were ball of Si_3N_4 and steel 16MnCr5-EN 10084-94, EN 84-70, results are shown in Figure 14 (right).

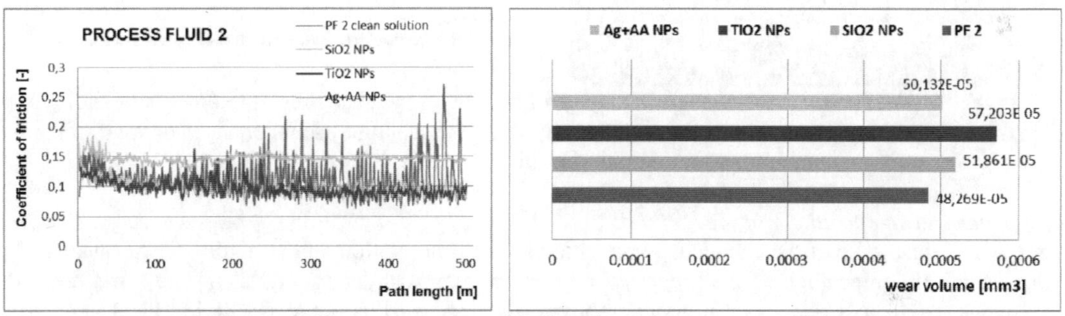

Figure 14: Coefficient of fiction and wear of balls (Si_3N_4) in the process fluid PF2 without/with nanoparticles.

Several important parameters for wear can be evaluated using the optical microscope, e.g. size of wear beads, shown in Figures 15 and 16 (top left of the figure).

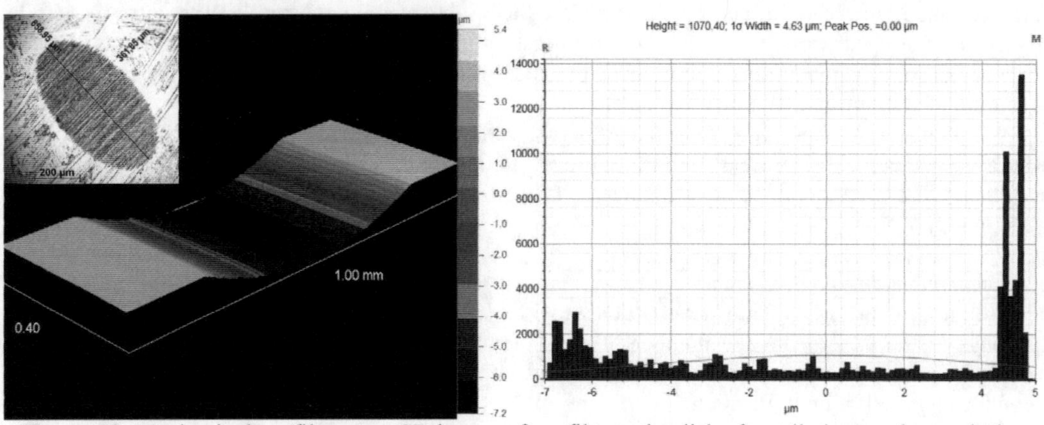

Figure 15. Mechanical profilometer: 3D image of profile on the disk after tribology - clean solution and optical microscopic images of wear of the ball.

Information on the volume of spalled material can be obtained using the Dektak XT ™ mechanical profilometer, Figures 15 and 16, which shows the wear when using the unmodified process fluid and process fluid with nanoparticles Silver nitrate ($AgNO_3$) and Ascorbic acid ($C_6H_8O_6$). The relevant software enables the measured values to be sorted using Gauss' law along the investigated profile and offers the possibility of sorting values in a histogram according to their frequency. The wear rate of the friction pair of two parts provides information on the friction properties of the media used under specific conditions (Figure 14 right - wear of ball Si_3N_4 without/with nanoparticles).

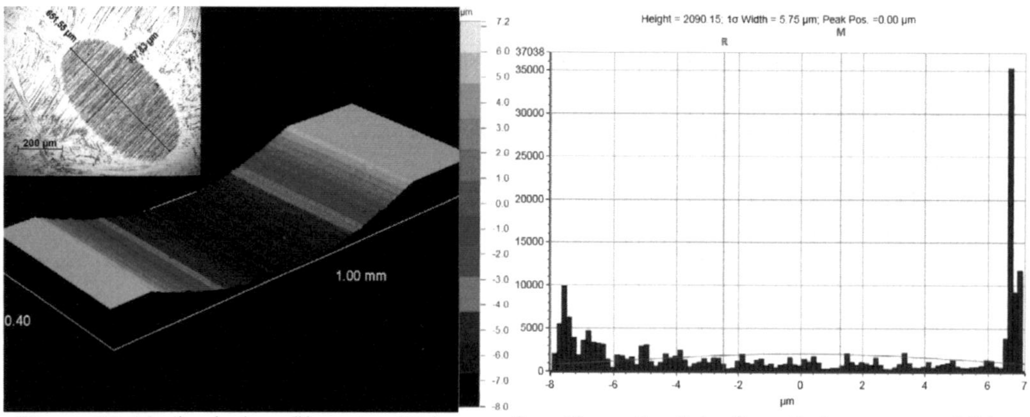

Figure 16. Mechanical profilometer: 3D image of profile on the disk after tribology - nanoadditive Silver nitrate ($AgNO_3$) and Ascorbic acid ($C_6H_8O_6$) and optical microscopic images of wear of the ball.

3.3. Chemical composition of process fluids

In the experimental part the chemical composition of the process fluids PF 1 and PF 2 was evaluated. The chemical composition of the process fluid concentrates was determined using the static method HS-GS-MS (Head Space-Gas Chromatography-Mass Spectrometry). Process fluid (PF 1) consists mainly of oxygenated organic compound types of glycols and higher alcohols, which are not part of the process fluid PF 2.

PF 1 contains methylene chloride, which is toxic to the human body and has been proven to have carcinogenic effects; it is also associated with a number of risks threatening human health. Dichloromethane is metabolized in the body to form carbon monoxide, consequently increasing the carboxyhemoglobin level in the blood and thus reducing the capacity of the blood to carry oxygen to induce respiratory distress, headache and dizziness. Other hazardous materials, components of PF1, are 1.4 - dioxane, dimethylformamide, trimethylsilanol and octametyhcyclotetrasiloxan belonging to the group of substances with proven toxic effects.

PF 2 is mainly composed of oxygenated organic compound types of glycols and higher alcohols, which are not part of the process fluid PF 1. From a toxicological point of view, PF 2 does not contain substances that have carcinogenic or mutagenic properties and causes no major negative impact on the human body.

3.4. Evaluation of biological tests

The addition of SiO_2 and TiO_2 nanoparticles to PF1 caused a reduction in cell number, even after 120 days of the experiment the number of cells did not change significantly, while in the second sample PF2 the number of cells increased. SiO_2 and TiO_2 nanoparticles in PF 1 have a long-term effect on reducing bacterial multiplication.

The addition of $AgNO_3$ nanoparticles into the process fluids caused a reduction in cell numbers even after 70 days of the experiment and the number of cells further decreased. $AgNO_3$ nanoparticles have a long-term effect on reducing bacterial multiplication.

The addition of ascorbic acid into the sample causes additional inhibition of the bacterial strain of *E. Coli*. The most significant biocidal effect of $AgNO_3$ was observed in PF 2. After two months of exposure the effect of the $AgNO_3$ nanoparticles in the studied samples was still observed resulting in reduced respiratory activity compared to the control sample (23% of the *E. Coli* strain).

4. Discussion

The use of natural fillers in the form of nanoparticles is highly efficient due to their high chemical and biological activity. The use of nanoparticles is one method for inhibiting bacteria and improving the biological, chemical and technological stability of the fluids.

Tribological tests were conducted in both process fluids under the same conditions. It has been shown that the addition of silver nitrate and ascorbic acid in a 5% aqueous solution of PF 1 has a positive effect on reducing the coefficient of friction during the tribological process (Figure 11). Adding nanoadditives in the form of nanoparticles leads to a faster stabilization of the tribological process under certain experimental conditions, which is a very important factor when machining. The rate of wear of the friction pair after adding nanoadditives to the solution PF 1 did not lead to significant changes.

For process fluid 2 the values of the coefficient of friction using a 5% solution with the addition nanoadditives TiO_2, SiO_2 and silver nitrate with ascorbic acid had no effect on the tribological properties (Figure 14). Adding nanoparticles of silver nitrate with ascorbic acid resulted in a significant reduction in the depth of material being removed, and furthermore to an increase in the groove width. For nanoadditives TiO_2 and SiO_2 it resulted in a decrease of groove width, even more significantly when using the solution of TiO_2 nanoparticles.

5. Conclusion

The tribological tests (Figure 11 and Figure 14) show that the addition of nanofillers SiO_2 and TiO_2 has no negative effect on the coefficient of friction during the tribological process. For PF 1 it resulted in a stabilization of the values of friction coefficient using a 5% solution with the addition of nanoadditives SiO_2, TiO_2 and silver nitrate with ascorbic acid. In process fluid 2 the values of the coefficient of friction using a 5% solution with the addition nanoadditives TiO_2, SiO_2 and silver nitrate with ascorbic acid had no effect on the tribological properties. The measured values are close to the values registered in the tribology of the pure 5% solution.

The biological experiments demonstrate the positive impact of the use of nanoadditives. Suppression of bacterial proliferation was achieved by using nano - SiO_2 35 ÷ 69%, nano-silver nitrate 60 ÷ 90% and nano - silver nitrate with ascorbic acid and 70 ÷ 90%, where significant toxic effects were observed in the bacterial strain of *E. Coli*.

The manufacturing process with process fluids has to be in accordance with European Union requirements that are specifically focused on environmental and health and safety standards. The addition of natural nanoadditives is one possible way of improving the PF properties and compliance with EU requirements.

Acknowledgement

The paper was supported in part by the project OPR&DI "Innovative products and environmental technologies", registration number CZ.1.05/3.1.00/14.0306 which is funded by the European Regional Development Fund through the Operational Programme Research and Development for Innovation. The results of this project LO1201 were obtained through the financial support of the Ministry of Education, Youth and Sports in the framework of the targeted support of the "National Programme for Sustainability I" and the OPR&DI project Centre for Nanomaterials, Advanced Technologies and Innovation CZ.1.05/2.1.00/01.0005.

References

[1] Hauk V and H Behnken 1997 Structural and residual stress analysis by nondestructive methods: evaluation, application, assessment. In: Materials science and technology (Boca Raton, Fla.), (New York: Elsevier) pp 640

[2] Humar A 2008 Materials for cutting tools. Praha: MM publishing, ISBN 978-80-254-2250-2, pp 235

[3] Holmberg K and A Matthews 2009 Coatings tribology: properties, mechanisms, techniques and applications in surface engineering. 2nd ed. Boston: Elsevier Science, Tribology and interface engineering, Series 56, ISBN 978-044-4527-509, pp 560

[4] Dichev D, Koev H, Bakalova T and Louda P 2014 *Measurement Science Review* **14(4)** 183-189

[5] Dugin A, Votocek J and Popov A 2014 *Manufacturing Technology* **14(2)** 149-153

[6] Dugin A and L Volesky 2014 *Manufacturing Technology* **14(3)** 290-296

[7] Bakalova T, Louda P, Volesky L and Andrsova Z 2014 *Manufacturing Technology* **14(3)** 256-261

[8] Dobes P 2013 Health and safety aspects of process fluids and their application, Techmagazin, http://www.techmagazin.cz/1135

[9] Nemec P 2013 Gentle process fluids at machining of metals, MM Industrial Spectrum 72-73

[10] Xiang L, Gao C, Wang Y, Pan Z, Hu D 2014 *Particuology* **17** 136-144

[11] Debnath S, Reddy M M, Yi Q S 2014 *Journal of Cleaner Production* **83** 33-47

[12] Chan C Y, Lee W B, Wang H 2013 *International Journal of Machine Tools and Manufacture* **73** 62-70

[13] Sayuti M A, Sarhan A D, Salem F 2014 *Journal of Cleaner Production* **67** 265-276

[14] Bakalova T, Louda P, Volesky L and Křiklavova L 2014 The impact of natural nanoaditiv on the tribological and chemical properties of process fluids. In: Proceedings: 8th International Conference on Tribology, 30thOct.-1stNov.2014, Sinaia, Romania. Ploesti: Petroleum-Gas University of Ploiesti Publishing House, ISBN 978-973-719-570-8, pp 143-148

[15] Holmeg K and Matthews A 1998 Coating tribology – properties, techniques and applications in surface engineering (Amsterdam: Elsevier)

[16] Burakowki T and T Wierzchon 1999 Surface engineering of metals (New York: CRC Press) ISBN 0–8493–8225–4

[17] Hodkova S 2004 Evaluation of mechanical properties of thermally sprayed coatings, [dissertation thesis], ZCU, Plzen

[18] Novotna K 2004 Tribological properties of thermally sprayed coatings, [dissertation thesis], ZCU Plzen

[19] Andrsova Z, Kejzlar P, Volesky L 2013 *Fine Mechanics and Optics* **1** 21-24

[20] Serra J 1983 Image Analysis and Mathematical Morphology (Orlando, FL, USA: Academic Press, Inc.) ISBN: 0126372403

[21] Kejzlar P, Macajova E and Kratochvil P 2013 *Fine Mechanics and Optics* **1** 25-27

MicroTherm'2015 and SENM'2015 IOP Publishing
Journal of Physics: Conference Series **709** (2016) 012002 doi:10.1088/1742-6596/709/1/012002

Polyurethane foams obtained from residues of PET manufacturing and modified with carbon nanotubes

L Stiebra[1, 2], U Cabulis[1] and M Knite[2]

[1]Latvian State Institute of Wood Chemistry, 27 Dzerbenes Street, Riga, Latvia
[2]Institute of Technical Physics, Riga Technical University, 3/7 Paula Valdena Street, Riga, Latvia

E-mail: laurastiebra@gmail.com

Abstract. In this work we report the preparation of rigid microcellular polyurethane/carbon nanotube nanocomposites with different CNT loadings (0.09-0.46%) and various isocyanate indexes (110-260). Water was used as a blowing agent for samples. Density of all obtained samples – 200 ± 10 kg/m^3. Electrical properties, as well as heat conductivity, cellular structure and mechanical properties of these nanocomposites were investigated.

1. Introduction

Polyurethanes (PUs) are segmented copolymers consisting of soft and hard segments. The first provide elastomeric properties, while second - dimensional stability. Neat PU foams were used as lightweight construction and isolating materials for various applications. To obtain materials with whole new set of properties, PU matrix can be modified by adding different reinforcing agents (organic or inorganic nanoparticles, nanofibers, nanoplatelets, etc.) [1, 2]. Carbon nanotubes (CNTs) have recently attracted a great deal of interest due to their high aspect ratio and surface area. If van der Waals forces between nanotubes are overpassed, they can be dispersed in polymer composites, CNTs can form conductive paths and the final composite presents improved electric, thermal and mechanical properties [1-5]. In order to enhance the electrical properties of polymer nanocomposite, it can be foamed – in this case nanoparticles (or nanotubes) are better distributed due to the use of a physical blowing agent [6]. The studies show that nano/microcellular polypropylene/multi walled CNT composites exhibits higher electrical conductivity, lower electrical percolation, higher dielectric permittivity and lower dielectric loss [6]. Carbon particles introduced into rigid PU foams can enhance material thermal insulating properties – cell walls and edges absorb increased amount of infrared radiation. Such additives also can act as fire retardants [3].

At relatively low CNTs concentrations and under specific conditions, a three-dimensional electrically conductive network may be formed in the polymer nanocomposite. It is achieved thanks to a physical contact between the dispersed conductive nanoparticles. This dramatically increased electrical conductivity is known as the electrical percolation threshold [4, 7]. Electrically conductive thermosetting nanocomposites have been obtained with extremely low CNT percolation thresholds (~0.0025 wt% CNT) and electrical conductivities as high as 2 S/m for CNT contents as low as 1 wt% [8].

In this work rigid PU/CNT composite foams were obtained. The influence on final composites of the CNT amount together with NCO index was determined. Mechanical properties, piezoresistivity as well as heat and electrical conductivity were investigated. The potential of the application of such composites in pressure sensors in automotive engineering were investigated.

2. Materials and methods

2.1. Materials

For preparation of PU foams as a main ingredient the polyether polyol Neopolyol® produced by Neo Group (hydroxyl number 370÷380 mg KOH/g, viscosity 3000÷4000 mPa·s, acid number <1 mg KOH/g, water content ≤0.1 %) was used. It is obtained in the transesterification of PET by diethylene glycol in the presence of functional additives [9]. Lupranol®3422 (hydroxyl number 490 mg KOH/g, viscosity 22750 mPa·s, water content ≤ 0.1 %) in the mixture was used as a functional polyol, Quadrol® (hydroxyl number 770 mg KOH/g, viscosity 52000 mPa·s, water content ≤0.1 %) - as cross-linking agent. Both are commercially available polyols from BASF. Distilled water was added as chemical blowing agent, while potassium acetate as trimerization catalyst and, tris(chloropropyl)phosphate as flame retardant. Surfactant Niax silicone L-6915 LV was purchased from Momentive and catalyst PC CAT®NP 10 from Air Products. Polyisocyanate PMDI was obtained from BASF. In order to improve PU matrix conductivity, multi walled CNTs NC7000 by Nanocyl™ were used (diameter 9.5 nm, length 1.5 μm, volume resistivity 10^{-4} Ω·m).

2.2. Samples preparation

First CNTs in different amounts (0.09 to 0.46 wt%) were incorporated and dispersed in polyol using high shear mixer (Silverson L5M-A) with a square hole high shear screen. The mixture was stirred for 30 minutes at 10000 rpm. In order to avoid undesirable reactions, which can happen because of friction forces and associated heating-up, the suspension was cooled in water bath (the temperature of 100°C was never exceeded).

After the catalyst (potassium acetate), blowing agent and surfactant were mixed with polyol and CNT blend, isocyanate in 4 various proportions was added. The suspension was stirred for approximately 10 seconds and poured in a preheated closed stainless steel form with dimensions 150 x 150 x 50 mm. Afterwards the mould was closed and placed into the oven at 50 °C for at least 2 hours to cure. The precise formulation in parts by weight is summarized in Table 1 [10].

Table 1. The base formulation of PU foams.

Component	Parts by weight
Neopolyol 380	78.0
Lupranol® 3422	18.0
Quadrol®	4.0
TCPP	20.0
H_2O	1.0
PC CAT® NP 10	1.9
Niax silicone L 6915 LV	2.0
Potassium acetate	1.6
NC7000	0.0 – 1.2
PMDI	130-307

2.3 Characterization

The density of samples was determined using samples intended for mechanical tests. The morphology of foam samples were characterized using SEM Tescan TS 5136 MM with secondary electron detector and VEGA TC computer software. Before taking SEM images, foam samples were cut into pieces of approximately 10 x 10 x 2 mm using scalpel blade and glued on aluminium specimen tables. To provide better electron conductivity, samples were coated with gold by sputtering using sputter coater Emitech K550X. In order to determine the average size of the foam cells, foam cell size distribution and anisotropy, cell diameters in both axial and transverse directions were measured.

To determine strength and modulus of the elasticity, compression tests were performed. In this aim, Zwick/Roell Z100 testing equipment was used. Samples for compression tests were cut parallel to the foam growth direction. The results are the average of the measurements made on 5 various foam samples. It was done accordingly to the standard EN ISO 844:2009.

Heat conductivity was determined with Linseis HFM 200 thermal analyser. The size of samples 195 x 195 x 50 mm. Temperature of the top plate was 20°C, temperature of the bottom plate was 0°C.

Volume resistivity measurements and necessary calculations were carried out on the basis of the standard ASTM D257. Picoammeter Keithley Model 6487 together with Resistivity Test Fixture Model 8009 was used.

Piezoresistivity was measured using test equipment consisting of Zwick/Roell Z2.5 compression force tester and Agilent 34970A Data Logger 3. The samples were 20.0 ± 0.5 mm in diameter and 10 ± 1 mm thick. To ensure a good contact with the electrodes, specimens were coated with silver paste and dried at room temperature for 24 hours. Samples were compressed to 0.1 MPa with speed 0.005 MPa/s and then relaxed for 50 s.

3. Results and discussion

In order to control the crosslinking density and the morphology of the final nanocomposite material, various NCO and OH group (NCO/OH) ratios and various concentrations of CNTs were used. The density of all PU foams was controlled by the foaming in a closed mould (200 ± 10 kg/m^3).

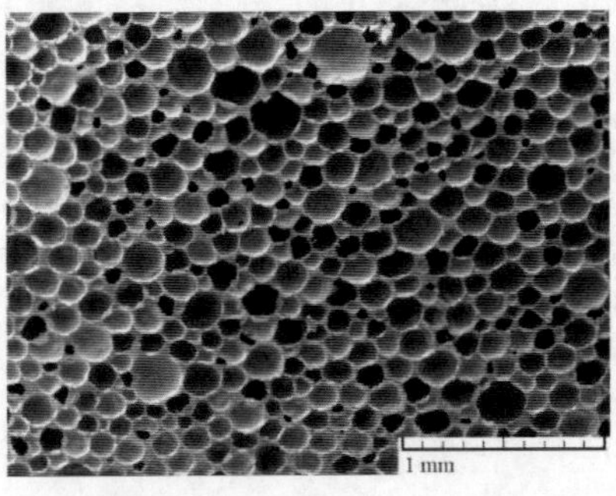

Figure 1. SEM micrographs of rigid PU foams with NCO index 160 unfilled (no CNTs present).

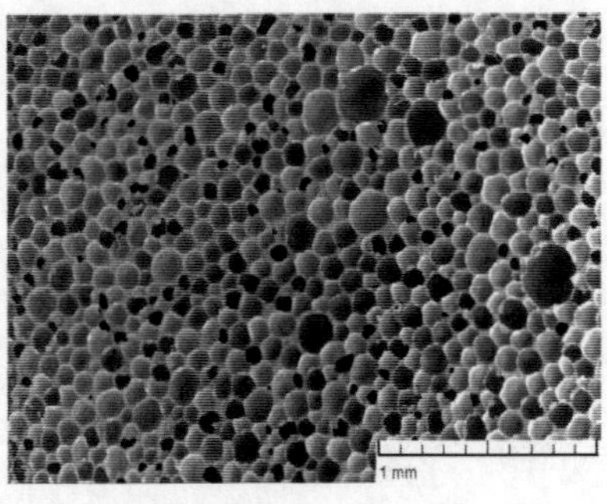

Figure 2. SEM micrographs of rigid PU foams with NCO index 160 filled with 0.12 wt% CNTs.

The morphology of rigid PU/CNT foams is observed by SEM. Figure 1 shows the microstructure of unfilled PU foam (no CNTs) and Figure 2 - PU foam filled with CNTs (0.12 wt%) (NCO/OH ratio - 160). The results show that with the increasing the amount of CNTs, the average size of foam cells decreases. It indicates that in small amounts of CNTs, they act as nucleation centres and produces finer cell structure and higher cell density. Overreaching 0.15 wt% of CNTs, the cell size increases in both axial and transverse directions (figure 3).

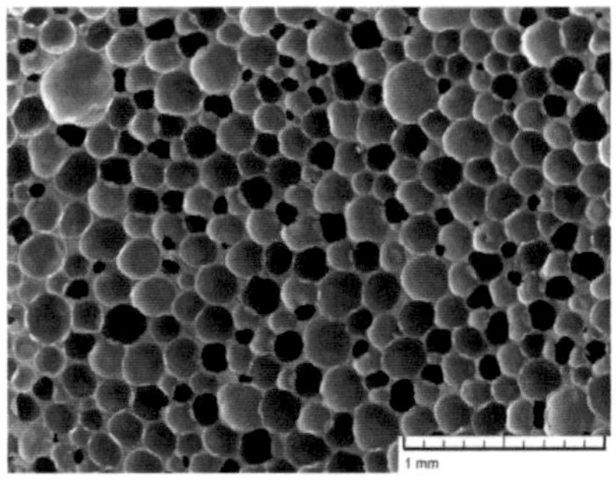

Figure 3. SEM micrographs of rigid PU foams with NCO index 160, filled with 0.21 wt% of CNTs.

Although the concentration of CNTs is, for example, 0.18 wt% in the final product, in polyol, in which additives were dispersed, CNTs constitute 1 wt%. Therefore the viscosity is high, dispersion difficult and foaming uneven. When increasing NCO/OH ratio, the foam cell size also increases. The main reason for such a result is that the same amount of surfactant is added in all cases, in order to provide a closed foam structure. All foam samples have an isotropic structure, which means that cells are close to an ideal circle.

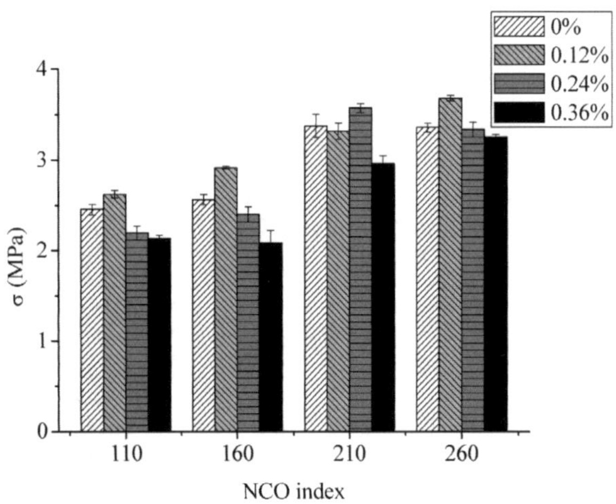

Figure 4. Compression strength of rigid PU foams with various NCO indexes and content of CNTs.

The obtained modulus and strength of PU foams from compression and tensile tests were normalized according to [11]. The results are given in Figure 4. It can be seen that when adding a small amount of CNTs, the compression strength becomes higher when compared to an unfilled sample. The same relation applies to the modulus of elasticity. If the CNTs content overreaches 0.15 wt%, both stiffness and strength decreases and these results correlate with the average foam cell sizes. With increasing NCO/OH ratio, the compressive strength increases.

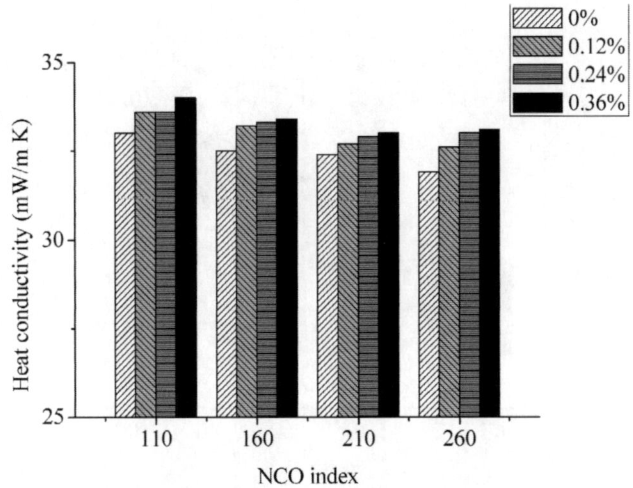

Figure 5. Heat conductivity of rigid PU foams with various NCO indexes and content of CNTs.

Heat conductivity of prepared composites is not dramatically affected neither by NCO index, neither by the concentration of dispersed CNTs (figure 5). However the trend is clear: when NCO index is increased, the thermal conductivity of the foams slightly decreases (33 to 31 mW/m·K).

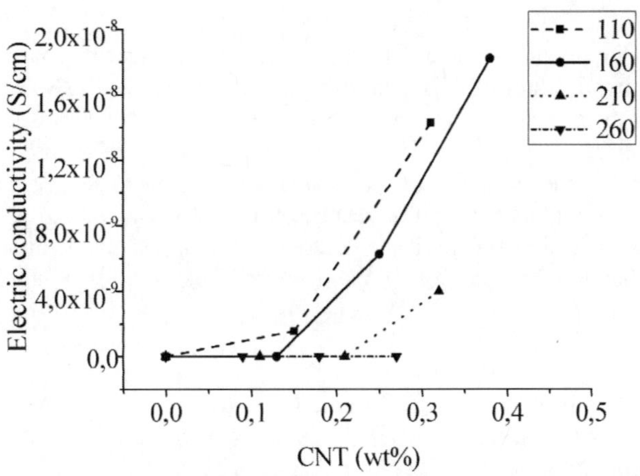

Figure 6. Volume electric conductivity of rigid PU foams.

From the Figure 6 it can be noticed that the volume electric conductivity for PU foams increases significantly, even if very small content of CNTs is added. In case of the NCO index of 110, adding only 0.3 wt% of CNTs, the conductivity improves by 5 to 6 orders of magnitude. CNTs and increased NCO index influence the electrical properties of PU foam in a negative manner – the conductivity decreases. The electric conductivity of foams with different NCO indexes and without additives remains the same (around 0 S/cm).

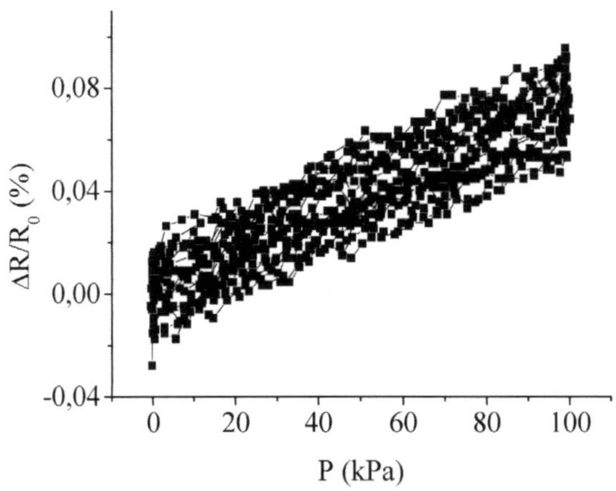

Figure 6. Piezoresistive sensitivity of rigid PU/CNT composite with 0.15 wt% of CNTs and the isocyanate index of 110.

All tested samples display positive pressure coefficient of resistance – by increasing the compressive force, the resistivity of composites increases (figure 5). The electrical resistance after several compressive loading cycles increases and does not return to its initial value. Thus it can be concluded that CNTs introduced in the nanocomposite under pressure do not form conductive networks – when the compressive force is applied, the distance between nanotubes grows.

4. Conclusions

The PU/CNT nanocomposite foams were obtained and their piezoelectric characteristics were tested. To produce PU foams, residues of PET manufacturing were used and CNTs were dispersed in liquid polyol using high shear mixer. Electrical conductivity for PU/CNT foams increases – with an increasing concentration of CNTs, but it is very difficult to effectively introduce more than 0.15 wt% of CNTs because of the viscosity of PU foams forming polyols. The most significant result is that electrical resistivity reduction can be reached if the NCO index is low. PU/CNT foam composites demonstrate the pressure sensitivity, therefore the obtained nanocomposite materials can find a potential application in automotive engineering, but additional studies should be performed in order to prove it.

5. References

[1] Zhang L, Yilmaz E D, Schjødt-Thomsen J, Rauhe J Chr and Pyrz R 2011 MWNT reinforced polyurethane foam: processing, characterization and modelling of mechanical properties. *Compos. Sci. Technol.* **71** 877-84

[2] Velayutham T S, Ahmad A B, Abd Majid W H and Seng Neon G 2008 Electrical properties of polyurethane graphite composites *Jurnal Fizik Malaysia* **29** 25-8

[3] Xu X-B, Li Z-M, Shi L, Bian X-C and Xiang Z-D 2007 Ultralight conductive carbon-nanotube-polymer composite *Small* **3(3)** 408-11

[4] Antuanes M and Velasco J I 2014 Multifunctional polymer foams with carbon nanoparticles *Prog. Polym. Sci.* **39** 486-509

[5] Thomassin J-M, Jérôme C, Pardoen T, Bailly C, Huynen I and Detrembleur C 2013 Polymer/carbon based composites as electromagnetic interference (EMI) shielding materials *Mater. Sci. Eng.* **74** 211-32

[6] Ameli A, Nofar M, Park C B, Pötschke P and Rizvi G 2014 Polypropylene/carbon nanotube nano/microcellular structures with high dielectric permittivity, low dielectric loss, and low percolation treshold *Carbon* **71** 206-27
[7] Kirkpatrick S 1973 Percolation and conduction *Rev. Mod. Phys.* **45** 574-88
[8] Sandler J K W, Kirk J E, Shaffer M S P and Windle A H 2003 Ultra-low electrical percolation threshold in carbon-nanotube-epoxy composites *Polymer* 44 5893-9
[9] Makuska R and Vitkauskiene I 2008 Glycolysis of industrial poly(ethylene terephthalate) waste directed to bis(hydroxyethylene) terephthalate and aromatic polyester polyols *Chemija* **19** 29-34
[10] Kirpluks M, Cabulis U, Ivdre A, Kuranska M, Zieleniewska M, Auguscik M 2016 Mechanical and thermal properties of high density rigid polyurethane foams from renewable resources *J. Renew. Mater.* **4(1)** 86-100
[11] Hawkins M C, O'Toole B J and Jackovich D 2005 Cell morphology and mechanical properties of rigid polyurethane foam *J. Cell. Plast.* **41** 267-85

Acknowledgements
The financial support of Latvian National Research Programme IMIS2 and the FP7 project EVOLUTION under grant agreement No.314744 are greatly acknowledged.

Utilizing of inner porous structure in injection moulds for application of special cooling method

M Seidl[1], J Bobek[1], J Šafka[2], J Habr[1], I Nováková[3] and L Běhálek[1]
[1]Department of engineering technology, Technical University of Liberec, Czech Rep.
[2]Laboratory of prototype technologies and processes, Technical University of Liberec, Czech Rep.

E-mail: martin.seidl1@tul.cz

Abstract. The article is focused on impact evaluation of controlled inner structure of production tools and new cooling method on regulation of thermal processes for injection moulding technology. The mould inserts with porous structure were cooled by means of liquid CO_2 which is very progressive cooling method and enables very fast and intensive heat transfer among the plastic product, the production tool and cooling medium. The inserts were created using rapid prototype technology (DLSM) and they had a bi-component structure consisting of thin compact surface layer and defined porous inner structure of open cell character where liquid CO_2 was flowing through. This analyse includes the evaluation of cooling efficiency for different inner structures and different time profiles for dosing of liquid CO_2 into the porous structure. The thermal processes were monitored using thermocouples and IR thermal analyse of product surface and experimental device. Intensive heat removal influenced also the final structure and the shape and dimensional accuracy of the moulded parts that were made of semi-crystalline polymer. The range of final impacts of using intensive cooling method on the plastic parts was defined by DSC and dimensional analyses.

1. Introduction

The injection moulds belong among the main aspects that affect the quality of the production technology (injection moulding). The production tools include tempering systems that control the thermal processes in the mould and thereby directly influence the final properties and dimensional stability of the products. Conventional cooling methods may not be able to provide homogeneous heat transfer from the entire volume of the geometrically complex part. To increase the efficiency of the cooling process many unconventional ways of cooling were developed (vortex tube, heat pipe, Contura system etc.). These cooling systems include cooling by liquid technical gases as well. In this study liquid carbon dioxide was used for cooling of modified shape inserts in injection mould. The shape inserts were adapted for this purpose. Liquid CO_2 was spread in open-cell structure that surrounded shape cavity of injection mould. The core of the bi-component core / shell structure was created by pore elements that increased the surface area available for heat transfer between the solid part and coolant (air, water, oil etc.). [1, 2]

2. Materials and methods

2.1 Cooling method employing liquid CO_2
Carbon dioxide is an inert, non-flammable, colourless gas and it is delivered in a liquid state for technician applications and must be compressed above 5,185 bars (45-65 bars in pressure bottles / bundles or 12-35 bars in isolated cylinders). Pressure drop of compressed CO_2 causes its rapid adiabatic expansion when leaving the bottle or cylinder and the Joule-Thomson effect generates reduction of temperature up to value of -78,9 °C at the ambient pressure. [3]
System assembled from tubes, capillary tubes, solenoid valves and control unit was completed to deliver "liquid" carbon dioxide to critical places inside manufacturing tools. This intensive cooling method was developed in the 90th as an output of cooperation between companies Linde Gas and IKS (Iserlohner Kunststoff Technologie GmbH) and this system was introduced under the trademark "Spot cooling". [4]

2.2 Bi-component injection mould insert
This article deals with high potential of porous metal materials in the field of intensive heat removal in combination with liquid carbon dioxide and their employing in the injection mould. The insert with bi-component core / shell structure was created with using of the DMLS (Direct metal laser sintering) technology.

2.3 Temperature fields analysis
The temperature field distribution on the part surface was observed after part removal from the mould in IR spectrum with using Flir ThermaCAM SC 660 and acquired images were further evaluated with using the software instruments of Flir company. Temperatures on the part surfaces allowed evaluating the efficiency of individual cooling variants.

2.4 Shape analysis
The progress of heat transfer from the product to the cooling system has direct impact on the shape and dimensional stability of the part. The shape stability analysis consisted of two steps. In the first stage the parts were digitized using a 3D optical measuring device Atos II and the data were then compared to the CAD model.

2.5 DSC analysis
The impact of different technological conditions on the morphology and mechanical properties of the product made of semicrystalline thermoplastic PE HD LITEN BB 29 (produced by Unipetrol) was monitored by differential scanning calorimetry. During solidification the macromolecules are formed into the structure that reflects the progress and intensity of heat removal. Each semicrystalline thermoplastic reacts characteristically to the heat removal and the cooling process should provide maximum level of macromolecule ordering and thus possibly the best mechanical properties. The test sample was subjected to a linear heating rate of 10 °C/min from a temperature of 25 °C to 150 °C. Tests were carried out according to the standard EN ISO 11357-1 with using of DSC 1 unit Stare System Mettler Toledo.

3. Results
In the first step of this study the dosing of liquid carbon dioxide into various structures was evaluated. For this purpose test samples were produced both either with the compact inner structure or bi-component (core / shell) test samples with the core created by fractal elements with size of 1 mm in the first variant and with size of 2,5 mm in the second variant. Total amount of injected cooling medium was 52,5 grams per one cycle and the batching of CO_2 followed adjusted profiles (three cooling modes). The first cooling mode denoted continuous injection of carbon dioxide for the duration of 5 seconds. In the second and third cooling mode the batching of CO_2 was divided into five

(solenoid valve was open 1 second and 1 second closed) respectively ten (solenoid valve was open 0.5 second and 0.5 second closed) pulses and the cooling cycle time had a duration of 10 seconds. Temperature distribution on the sample surfaces was monitored with using of two K thermocouples placed at locations situated geometrically to outfall of capillary tubes. The most effective cooling was observed when injecting the liquid CO_2 in the pulse mode with pulse duration of 0,5 s. The most uniform spread of liquid coolant was monitored inside the structure formed by fractal elements with size of 2,5 mm. Results from the first phase of the study were transferred to the production stage, where the evaluation of cooling efficiency for both cooling systems was performed. One shape insert was modified for the cooling by water that circulates in drilled tempering channels. Into this shape insert the expansion chambers were also drilled. The presence of expansion chambers allowed monitoring of the combined effect of simultaneous cooling in the conventional way and cooling with CO_2. The second shape insert had a bi-component structure (core / shell) with a core consisting of the fractal elements with size of 2,5 mm. Tempering water had a temperature of 40°C. Liquid CO_2 was injected into the shape insert in pulses with length 0,5 seconds. Distribution of temperature fields on the part surface was monitored at different cooling time (namely after 5, 10 and 25 seconds of cooling). The cooling time was set only on 5 and 10 seconds when using the liquid CO_2, see figure 1(left) and table 1. The produced parts were subjected to shape analyse and the results are shown in figure 1 (right). The influence investigation of the unconventional method of cooling on the morphology of produced parts was the last step of this analysis. For this purpose the DSC analysis was performed that revealed a level of macromolecular structure depending on the heat transfer progression, see table 2.

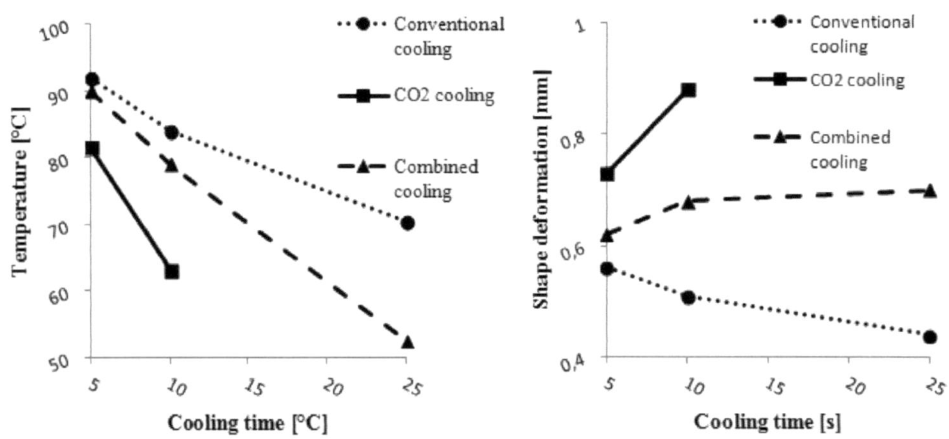

Figure 1. Graphical comparison of mean values of part surface temperatures (left) and comparison of part dimensional stability (right).

4. Discussion

In the first part of this analysis the effective dosage of liquid CO_2 into a production tool was determined. The most intensive heat removal was observed when injecting the liquid carbon dioxide in the pulse way into the porous structure created by the biggest analysed fractal elements. Based on these results mould insert was designed. During injection moulding of test samples the distribution of temperature fields on the surface of the products was monitored after their demoulding from the mould. The results summarized in table 1 imply that very uniform temperatures on the product surface were achieved when cooling in conventional way. The longer the product was cooled, the lower the temperatures on the part surface were. Cooling with using of liquid CO_2 was assumed to be more

Table 1. Maximum and minimum temperatures on the part surface.

		Cooling time		
		5 s	10 s	25 s
Conventional cooling	max.	93 °C	85 °C	72 °C
	min.	91 °C	83 °C	69 °C
CO_2 cooling	max.	79 °C	66 °C	
	min.	84 °C	60 °C	
Combined cooling	max.	93 °C	81 °C	54 °C
	min.	87 °C	77 °C	51 °C

intense. However temperature distribution on the part surface was not so uniform which arose from location of the capillary tubes that fed liquid CO_2 into the mould insert. Combined cooling showed again relatively even distribution of temperature fields on the part surfaces. Increasing cooling time of combined cooling method did not bring a positive effect on the uniformity of temperature distribution on the part surface. Different progress of solidification process had a direct impact on the dimensional stability of the product after its removal from the mould. Even temperature field indicated better dimensional stability of the part, see figure 1 (right). The lowest level of deformations was achieved when cooling in conventional way. The lower temperature was on the part surface after its removal from the mould, the greater dimensional stability of the product. On the contrary relatively large shape deformations appeared when application of liquid CO_2. With increasing cooling time the dimensional stability was even reduced. For combined cooling similar trends as for cooling with CO_2 were observed. With increasing cooling the part deformation increased as well. Measured values ranged between deformations obtained by individual cooling systems separately. Cooling with using of liquid CO_2 did not affect positively the morphology of the products, see table 2.

Table 2. DSC analyse of test samples.

	Cooling Time	ΔH	Onset of melting	Endpoint of melting
Pure PE	0 s	174.8 J/g	84 °C	142 °C
Conventional cooling	5 s	165.6 J/g	92 °C	146 °C
	10 s	163.1 J/g	90 °C	143 °C
	25 s	160.9 J/g	89 °C	142 °C
CO_2 cooling	5 s	159.2 J/g	86 °C	145 °C
	10 s	155.3 J/g	83 °C	144 °C
Combined cooling	5 s	163.6 J/g	89 °C	145 °C
	10 s	161.3 J/g	89 °C	143 °C
	25 s	159.8 J/g	87 °C	142 °C

The best arrangement of the macromolecules was achieved when cooling in conventional way in the duration of 5 seconds (the shortest analysed cooling time). Uniform morphological structure was created with regular formations of macromolecules which disintegrated when reaching relatively high temperature. Above this temperature the structure disintegrated relatively quickly. In the case of parts cooled with using of CO_2 less amount of heat was necessary to be supplied to disintegrate their morphological structure. Smaller macromolecular formations were disintegrated already at lower temperatures because the intermolecular forces were not strong. These facts indicated lower mechanical properties. With longer cooling time by means of liquid CO_2 the uniformity of macromolecular formations was decreasing.

5. Conclusions

Liquid CO_2 ensured very intense heat removal from the injection mould. From the viewpoint of uniformity of part cooling the application of liquid carbon dioxide did not obtain better results than the conventional cooling method. The location of capillary tubes had a significant impact on the progress of heat transfer from the plastic part. Considering high cooling potential of liquid carbon dioxide it is also necessary to accurately specify the amount of injected cooling medium to avoid the overcooling of the production tool. Suitable applications where liquid CO_2 could be employed the combined methods of cooling are where this intensive cooling works together with conventional cooling way. Liquid CO_2 ensures very fast heat removal from hot spot areas in the mould and conventional tempering system enables the maintaining the tool at desired operating temperature.

6. References

[1] Kennedy A 2012 Porous Metals and Metal Foams Made from Powders *Powder Metallurgy* DOI:10.5772/33060
[2] Qu Z, Xu H, Wang T, Tao W, Lu T 2011 Thermal Transport in Metallic Porous Media *Heat Transfer - Engineering Applications* DOI: 10.5772/27018
[3] Union engineering 2007 *Properties of carbon dioxide*
[4] Resgren U, Praller A 2004 Industrial gases in plastics processing *Kunststoffe* (9) 251-252 Retrieved from: http://www.kunststoffe-international.com/PE102988

Acknowledgment

The published results were obtained through the financial support of the project LO1201 that was created in the framework of the targeted support of the "National Programme for Sustainability I" and the OPR&DI project Centre for Nanomaterials, Advanced Technologies and Innovation CZ.1.05/2.1.00/01.0005.

Structural properties of H13 tool steel parts produced with use of selective laser melting technology

J Šafka[1], M Ackermann[1] and L Voleský[1]

[1] Technical University of Liberec, Institute for Nanomaterials, Advanced Technologies and Innovation, Studentská 2, Liberec, Czech Republic

E-mail: `jiri.safka@tul.cz, michal.ackermann@tul.cz, lukas.volesky@tul.cz`

Abstract. This paper deals with establishing of building parameters for 1.2344 (H13) tool steel processed using Selective Laser Melting (SLM) technology with layer thickness of 50 µm. In the first part of the work, testing matrix of models in the form of a cube with chamfered edge were built under various building parameters such as laser scanning speed and laser power. Resulting models were subjected to set of tests including measurement of surface roughness, inspection of inner structure with aid of Light Optical Microscopy and Scanning Electron Microscopy and evaluation of micro-hardness. These tests helped us to evaluate an influence of changes in building strategy to the properties of the resulting model. In the second part of the work, mechanical properties of the H13 steel were examined. For this purpose, the set of samples in the form of "dog bone" were printed under three different alignments towards the building plate and tested on universal testing machine. Mechanical testing of the samples should then reveal if the different orientation and thus different layering of the material somehow influence its mechanical properties. For this type of material, the producer provides the parameters for layer thickness of 30 µm only. Thus, our 50 µm building strategy brings shortening of the building time which is valuable especially for large models. Results of mechanical tests show slight variation in mechanical properties for various alignment of the sample.

1. Introduction

The 3D printers for processing of metal materials give an amazing possibility of sintering various types of metal powders [1, 2, 3]. The result is a metal piece which may involve the complex geometries of surfaces or the porous structures. Many "printed" parts are therefore impossible to be produced with conventional methods (machining, metal casting, etc.) or it would demand complex and expensive approaches. Important factors in the processing (melting) of these materials are laser parameters, grain size of the powder, etc. [4, 5, 6]. Laser parameters directly affect the final quality and also the accuracy of the final part [7, 8]. In the case that different thickness of the layer is used, different laser parameters must be set. In this paper, the optimal parameters for the H13 tool steel with layer thickness of 50 µm are searched.

The sintering process was carried out on the SLM 280HL (SLM Solution GmbH) machine. This device, together with its software, allows the complete setting of individual laser parameters for a given layer. The thickness of the layers can be set from 20 to 100 µm. It is important to find the optimum ratio between the thickness of the layers, surface quality, internal structure quality and the time of production of the part. Due to the fact that supplier of the machine provided us with 30 µm parameters only, we wanted to develop parameters for frequently used 50 µm thick layer as well. This layer thickness is valuable especially in the case of large models because it dramatically reduces their building time.

2. Materials and methods

2.1. Laser parameters for H13 tool steel with layer thickness of 50 µm

Process of finding proper parameters for H13 tool steel with layer thickness of 50 µm was based on building several testing series of a cube model with chamfered edge (Figure 1 and 2). Each sample was then subjected to set of tests which goal was to evaluate quality of its inner structure and surface roughness. Namely, these tests were:

- Measurement of surface roughness
- Metallography and Light Optical Microscopy (LOM)
- Scanning electron microscopy (SEM)
- Measurement of the micro-(nano) hardness

On the basis of these results, we were able to find such settings of laser power and laser speed which produces a model with solid core structure and low porosity. Due to the fact that border layers of the model are treated in a different manner, we had to find laser parameters for this part as well. Main goal for this part of the model is to produce lowest possible surface roughness.

Figure 1. CAD model of the sample.

Figure 2. Printed testing series.

Settings of laser power and laser speed for both volume part of the model and its border part were defined in the range shown in Table 1.

Table 1. Ranges of parameters for H13 tool steel testing series.

	Laser power [W]	Laser speed [mm/s]
Outer shell	50 – 150	340 – 540
Volume	125 – 225	400 – 600

As it can be seen in Figure 2, size of the testing matrix was 6x6 elements. In this testing matrix, upper left element is characterized with lowest value of laser power and speed whereas lower right element is built under highest values from the range shown in Table 1. All the other matrix elements are built with linearly extrapolated parameters between the border values.

2.2. Tensile tests of the specimens
Second part of the work dealt with mechanical properties of the parts which were printed under our settings. For this purpose, another set of samples in the form of "dog bone" were printed (Figure 3).

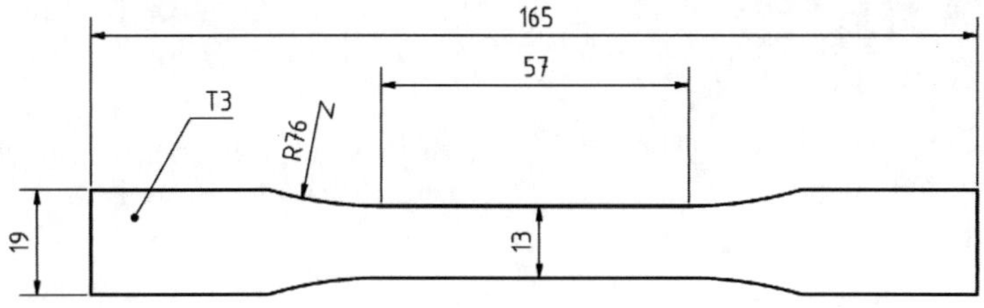

Figure 3. Tensile test specimen for evaluation of mechanical properties.

Apart from standard mechanical values such as ultimate tensile strength R_m of the material and its Young's modulus of elasticity E we were also interested if the different orientation of the model towards the building plate somehow influences mechanical properties of the material. Tensile test specimens were therefore printed in three different angles relatively to the building plate (Figure 4). Another interesting comparison of mechanical properties was done for the samples which were produced under our settings for 50 µm layer and the samples built with 30 µm layer thickness and settings provided by the manufacturer.

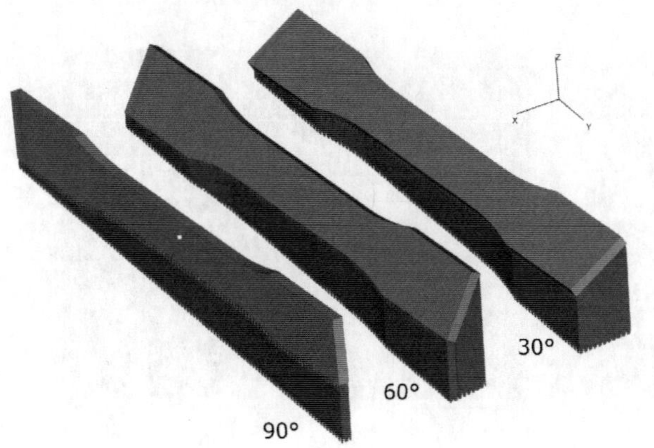

Figure 4. Three different positions of the specimen relatively to the building plate.

3. Results

3.1. Measurement of surface roughness
The surface roughness was measured to verify the quality of the border of the part and to measure the continuity of the individual layers. For the surface roughness measurement, the mechanical profilometer Bruser DeckTac XP with the roundness of the tip of 2 µm was used. On each of the 36 samples, the roughness was measured in three areas. Thanks to the results from roughness measurement it was possible to select several suitable samples for further analysis. Final measurement then consisted of evaluating real 3D surface topography of the last printed layer. In Figures 5 and 6 the 3D topography of the best and the worst surface of the printed samples is shown.

Figure 5. Specimen with the best resulting surface roughness

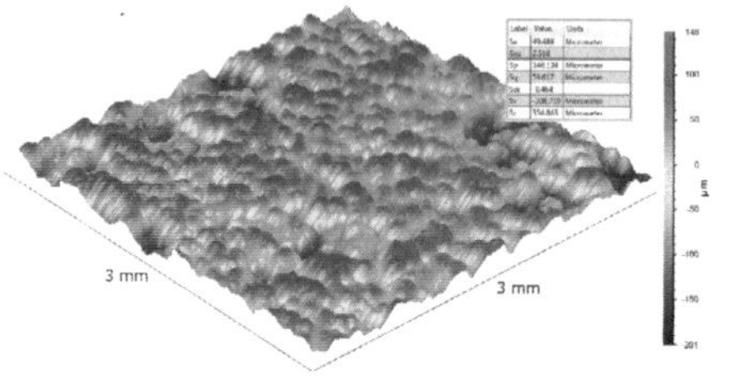

Figure 6. Specimen with the worst resulting surface roughness

3.2. Metallography and Light Optical Microscopy (LOM)

Main goal of this work was evaluation of material porosity and thus the quality of material's inner structure. Several devices were used for the sample preparation. Material cutting was done perpendicularly to sintered layers of the material on the precise metallographic saw Struers Secotom-50. After this process, the samples were immersed in conductive resin named PolyFast using the metallographic press Struers CitoPress-1. For subsequent grinding and polishing, the automatic metallographic polisher Struers Tegramin-25 was used. In order to accentuate the structure of the examined sample, etching was done by immersing the sample into Lorason solution for 15 s. Finally, scanning of structure was performed using a light optical microscope Carl Zeiss AXIO Imager M2 with an automatic shift in the mosaic mode with 100x magnification. In Figures 7 and 8 the resulting microstructure for a sample with minimum and maximum porosity, respectively, is shown.

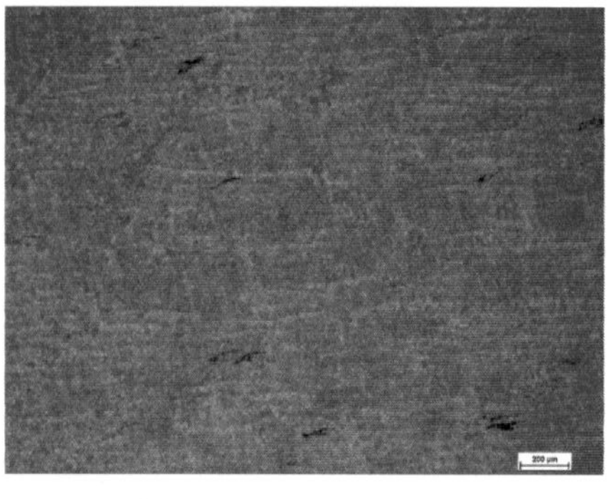

Figure 7. Specimen with minimum porosity

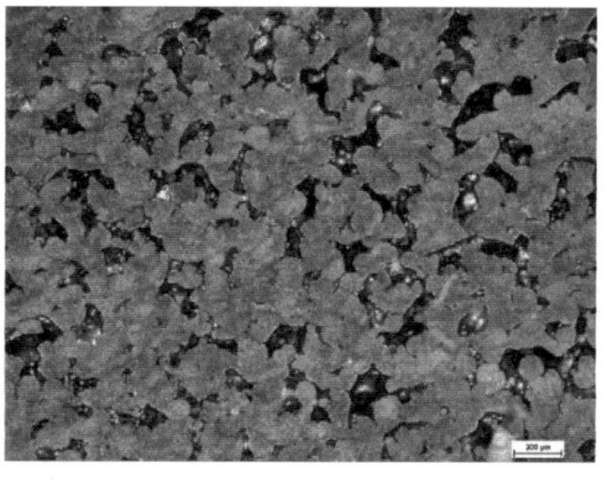

Figure 8. Specimen with maximum porosity

3.3. Scanning electron microscopy (SEM)

With use of scanning electron microscope Carl Zeiss Ultra Plus, we were able to investigate whether volume and outer shell of the part are correctly connected. Together with visual analysis, chemical micro-analysis was done by scanning electron microscope. This analysis is used to evaluate differences in chemical composition of volume and shell part of the model. Typical results for most of the specimens are shown in Figures 9 and 10. In the Figure 9, the border between volume and shell part of the specimen is clearly visible. In comparison with volume area, shell part is characterized with finer structure which is caused by double processing of the skin layer. After the implementation of chemical analysis, the possibility of contamination was eliminated as shown in Figure 10. Apart from standard alloying elements of H13 tool steel, there are no artificial elements present. On the other hand, there is a noticeable difference in the chemical composition of the layer and the substrate because the substrate contains larger amount of oxygen. This phenomenon is most probably caused by etching which causes corrosion of the material.

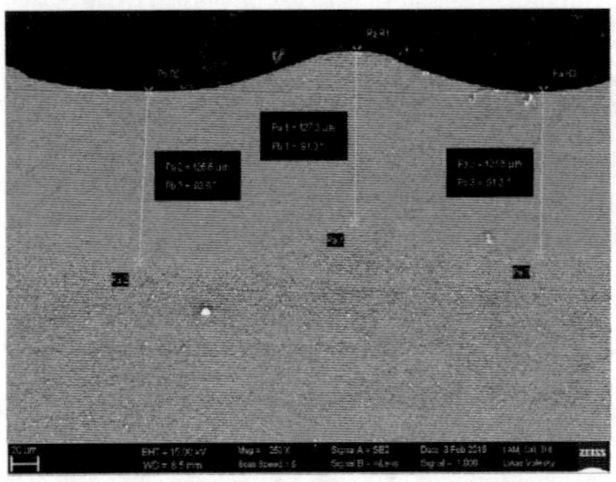

Figure 9. SEM picture of shell/volume connection

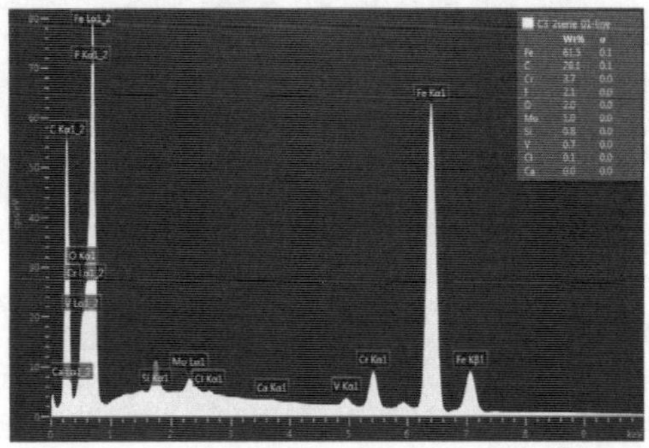

Figure 10. Result of micro-chemical analysis of shell/volume part

3.4. Measurement of the micro-(nano) hardness

This last test is another way to compare difference between properties of skin and volume part of the sample. In this case, the indentation test was performed with use of CSM Instruments Nano-Hardness device. In cross-section of the metallographic cut, four indentation tests were done into the both layer and the substrate material. Results of the tests are shown in Table 2.

Table 2. Results of micro-hardness measurement.

test ID		1	2	3	4	Average
Outer shell hardness	[HV]	745	738	747	748	**745**
Volume hardness	[HV]	620	626	575	628	**612**

3.5. Mechanical tests of H13 tool steel samples

After completion of the building process, the samples were cleaned and tested on the universal testing machine TIRA with the force sensor with force range of 100 kN. Each sample was clamped into specially shaped jaws and prestressed to a value of 1 kN. At the end of prestressing part, the mechanical extensometer with initial length of $l_0 = 30$ mm between the pins was attached to the sample. Concerning setup, tests were done under position control with loading rate of 5 mm/min until rupture of the specimen occurred (Figure 11). In the post-processing part of the tensile tests, the resulting channels of force and elongation from the extensometer were converted to the values of engineering stress and strain with respect to dimensions of the sample. Final tensile test diagrams were then used for comparison of all the desired samples.

Figure 12 shows tensile test diagrams of the samples which were built with the 50 μm building strategy and under different angle relatively to building plate. As apparent from the figure, at least two samples were tested for each angle. One can note that the samples built under angle of

Figure 11. Tensile test of the H13 tool steel specimen.

30° relatively to building plate shows highest values of ultimate tensile strength. Second set of tensile test diagrams (Figure 13), show comparison of the data for samples built under constant angle of 90° and with use of 50 μm and the original 30 μm building strategy. This graph clearly shows that no major change in mechanical properties is present for the samples built with our parameters for 50 μm layer and the samples built with original strategy for 30 μm layer.

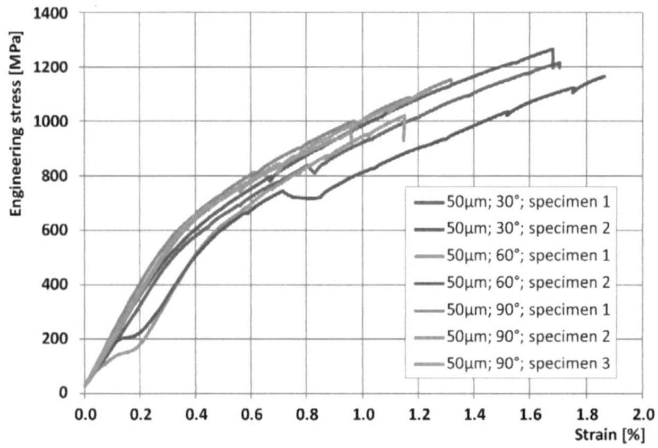

Figure 12. Tensile tests of the samples built under different angles relatively to building plate.

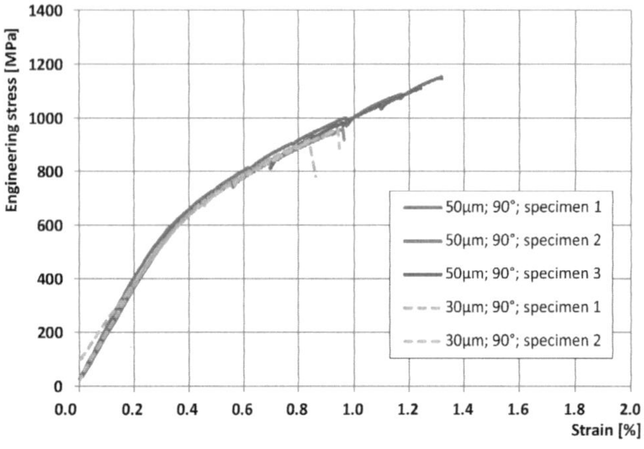

Figure 13. Comparison of results for samples printed with 50 μm and 30 μm layer thickness.

4. Discussion

First part of the work was aimed to find laser parameters for H13 tool steel powder and 50 μm thick layer. Several testing matrices of samples were printed and examined using variety of material tests. As it can be seen in Section 3, our most successful specimen has mean surface roughness of 24 μm and inner structure with minimum porosity. For comparison, Badrossamay et. al [9] reports average value of surface roughness to be 26.9 ± 3.1 μm for scanning parameters of 110 W laser power and 15 mm/s scanning speed. Further testing then showed that even though the machine is used for several powders, no contamination of the material with another alloying elements was found. Nano-indentation of the volume and skin part of the specimen also showed that there is a difference in hardness of these two parts. While skin part of the sample has average hardness of 745 HV (62 HRC), the hardness of the core is slightly lower – 612 HV (56 HRC). As it was mentioned before, this fact is most probably caused by multiple laser exposure of skin layer.

In the second part we tested mechanical properties of the samples which were printed with newly found parameters. Ultimate tensile strength of the samples varies between 1000 and 1200 MPa. Different orientation of the sample somewhat affected mechanical properties of the material. This effect can be caused by different layering of the material which leads to variation of mechanical properties in certain direction. Similar observation is described in the work of Holzweissig et al. [10]. The authors performed tensile test of the specimens which were printed perpendicular and parallel to the building direction. Reported ultimate tensile stress values for H13 tool steel varies between 1550 and 1650 MPa and 1150 and 1175 MPa, respectively.

5. Conclusions

Results which were proposed in this paper show that we reached our goal, i.e. successful development of the parameters for H13 tool steel powder and the layer thickness of 50 μm. With these settings, it will be possible to build large models in shorter time while preserving dense inner structure of the model and its appropriate mechanical properties. Mechanical testing of the samples also shown that there is a difference in mechanical properties of the samples which were build under different orientation versus the building plate of the machine. This phenomenon will be studied more in detail in our further work.

Acknowledgments

The results of this project LO1201 were obtained through the financial support of the Ministry of Education, Youth and Sports in the framework of the targeted support of the "National Programme for Sustainability I" and the OPR&DI project Centre for Nanomaterials, Advanced Technologies and Innovation CZ.1.05/2.1.00/01.0005.

References

[1] Zhang B, Dembinski L and Coddet C 2013 *Mat. Sci. Eng. A-Struct.* **584** 21–31
[2] Feuerhahn F, Schulz A, Seefeld T and Vollertsen F 2013 *Phys. Proc.* **41** 843–848
[3] Spierings A, Levy G, Labhart L and Wegener K 2011 *Virt. Rap. Man.* 785–790
[4] Yan C, Hao L, Hussein A and Raymont D 2012 *Int. J. Mach. Tool. Manu.* **62** 32–38
[5] Yadroitsev I, Bertrand P and Smurov I 2007 *Appl. surf. sci.* **253** 8064–8069
[6] Kruth J P, Froyen L, Van Vaerenbergh J, Mercelis P, Rombouts M and Lauwers B 2004 *J. Mater. Process Tech.* **149** 616–622
[7] Kurzynowski T, Chlebus E, Kuźnicka B and Reiner J 2012 *SPIE LASE* (International Society for Optics and Photonics) pp 823914–823914
[8] Hauser C, Childs T, Taylor C, Badrossamay M, Akhtar S, Wright C, Youseffi M, Xie J, Fox P and O'Neill W 2003 *Proceedings of the 2003 Solid Freeform Fabrication Symposium*
[9] Badrossamay M and Childs T 2006 *Proc. SFF Symp., Austin, Texas, USA* pp 268–279
[10] Holzweissig M J, Taube A, Brenne F, Schaper M and Niendorf T 2015 *Metallurgical and Materials Transactions B* **46** 545–549

Laser processing for bevel termination of high voltage pn junction in SiC

A Kubiak[1], Ł Ruta[1], A Rosowski[2,3], P French[2]

[1] Department of Semiconductor and Optoelectronic Devices, Lodz University of Technology, Wolczanska 211/215, 90-924 Lodz, Poland
[2] General Engineering Research Institute (GERI), Liverpool John Moore University, Byrom Street, Liverpool, L3 3AF, United Kingdom
[3] SPI Lasers, 3 Wellington Park, Tollbar Way, Hedge End, Southampton, Hampshire SO30 2QU, UK

E-mail: andrzej.kubiak@p.lodz.pl, lukasz.ruta@p.lodz.pl, adam.rosowski@spilasers.com, p.french@ljmu.ac.uk

Abstract. Proper edge termination of the p-n junction in silicon carbide is a key requirement in the fabrication of discrete devices able to withstand high voltages in reverse polarization. Due to the hardness of SiC the creation of the bevel termination remains difficult using mechanical machining. The use of laser beam sources with medium wavelength (532 nm) gives new possibilities in the machining of the silicon carbide. The paper presents the fabrication of the bevel termination structure in SiC using a green DPSS laser equipped with scanner and dedicated rotating sample holder. Characterization of the resulting structures proves the high potential of the proposed approach.

1. Introduction

Silicon carbide (SiC) is a wide-bandgap semiconductor that offers many advantages over common silicon for power devices, as it can be doped to higher levels than silicon to achieve a given blocking voltage. High voltage devices require a proper edge termination to eliminate early breakdown caused by the edge electric field intensification. For SiC devices the critical electric field is about one order of magnitude higher than of silicon, therefore the stress on the passivating dielectric can be much higher for SiC devices at breakdown than that for Si devices. Moreover, the high temperature capability is one of the advantages of SiC over other semiconductors. To take advantage of this, fabrication technology of SiC high-field devices capable of operating at much higher temperatures than Si devices must be developed. Since the reliability of any dielectric degrades when the temperature increases, reducing of the surface and edge electric field becomes critical for SiC high-field devices. This problem may be addressed by the edge termination technique, which can substantially reduce the electric field at the edges of the high voltage SiC device and thus improve its reverse characteristic.

1.1. Bevel termination of p-n junction

The process of bevelling consists of removal of semiconductor material at the edges of device at a precisely defined angle. The bevelling of the edges has been demonstrated to enhance the breakdown voltage by reducing the electric field at the edges as compared to cut orthogonal to the surface [1]. An example of the power devices with bevel edge termination at the upper high voltage junction is presented in figure 1. Mechanical grinding or abrasive powder bombardment are used for bevelling in silicon devices, but for silicon carbide those techniques are not successful.

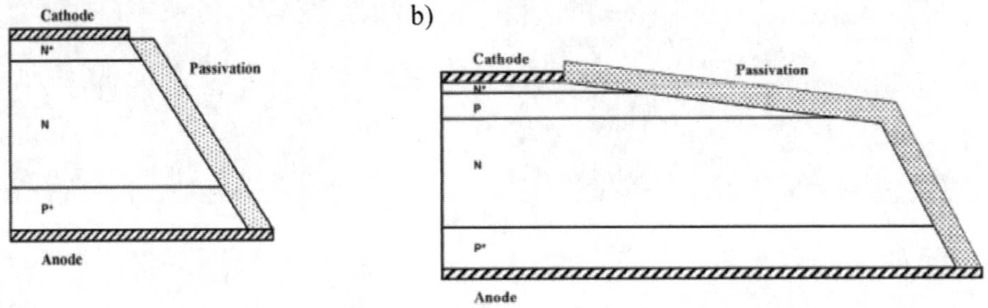

Figure 1. Cross section of the bevelled power devices: rectifier (a) and thyristor (b) [1].

1.2. Laser machining for bevelled SiC structures

Due to its hardness and smaller chips as compared to silicon ones, silicon carbide remains one of the most difficult material for mechanical treatment and thus fabrication of the bevelled surface by means of grinding may cause many troubles [2]. On the other hand, techniques based on the selective ICP etching of bevelled structures are quite complicated (multistep masking) and results in rough SiC surfaces [3]. Nowadays, the new possibility opens for maskless, non-contact bevel termination fabrication for SiC devices due to the laser technology development [4]. UV (335 nm) and green (532 nm) lasers are interesting for the ablation cutting of wide bandgap materials that have a low absorption coefficient or are nearly transparent at NIR (including 1064nm) wavelengths, such as GaAs, sapphire or glass [5,6]. The high photon energy of UV and green lasers plays also a significant role in laser machining of polymer materials such as polyimide. In addition to the absorption coefficient different ablation results are expected for ultraviolet, green and infrared lasers. In many cases, the heat diffusion length should not exceed the optical penetration depth, in order to obtain high quality microstructures with clean edges and smooth surfaces which are required for junction termination. As a result, many laser parameters, such as the frequency and duration of the laser pulses as well as scanning parameters needed to be considered too [7].

2. Experiment and results

Pure 4H-SiC substrate cut to 5x5 mm samples were used for laser machining. The green laser (532 nm, 6 W power output, output power variation <1%) equipped with beam scanner and dedicated rotating sample holder were employed to fabricate the bevelled structures. In the experiment the bevel angle adjusted by the holder position was set to 10°, 20°, 30° and 40° (±2°).

The images of the structures obtained by Zeiss Evo25 SEM are presented in figure 2. It can be seen that the bevelled surface remains well shaped and recurrent for all samples. However, for smaller laser beam angle (10°) it is difficult to keep the top surface undamaged and the interface between the bevelled and top surfaces become rounded (figure 2d). This effect is caused by insufficient precision in the available sample holder combined with the need of manual adjustment of the rotating sample according to laser beam focus position.

For the decreasing incident beam angle, the top surface of the device is also decreasing, but this effect is caused mainly due to the limited dimensions of the samples. To keep the active area of devices equal regardless of the bevelling angle, the samples would need to be bigger when smaller bevel angles are to be fabricated.

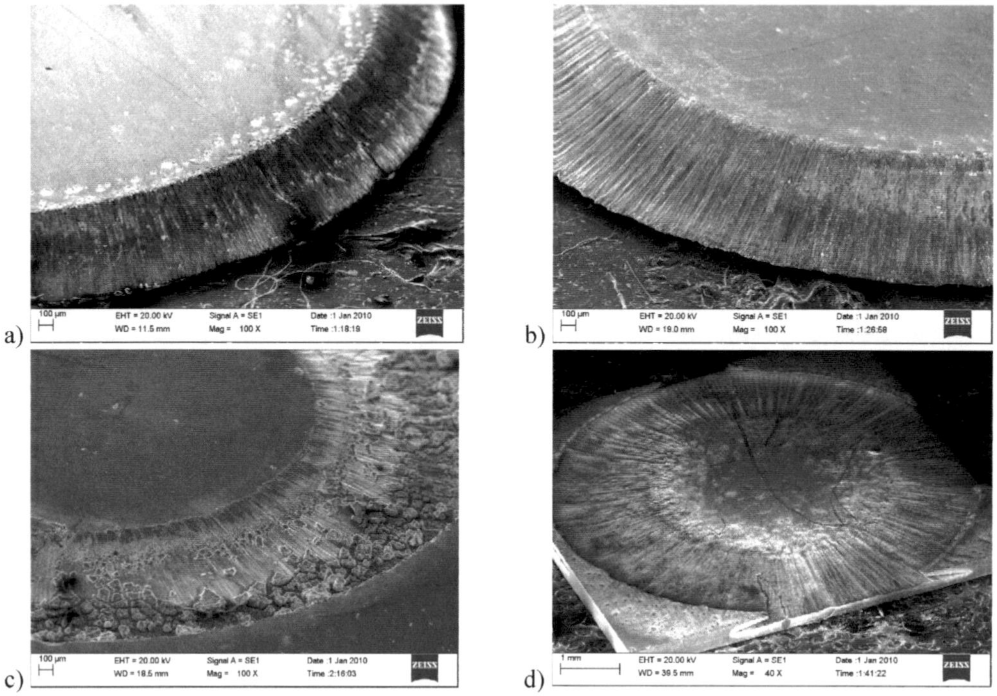

Figure 2. SEM images of the SiC laser cut structure with bevel angle 40° (a), 30° (b), 20° (c) and 10° (d).

In some circumstances the surface of the device may be accidentally damaged by the laser beam, as it can be seen in figure 2a. Therefore the additional protective layer covering the top surface of the structure would be beneficial. In case of complete device (e.g. diode), a rounded aluminium or nickel contact layer would act as a good protective layer since metals reflect laser beam nearly completely.

Due to very short laser pulses used during the machining process (35 ns) and low pulse repetition (31 kHz) straight lines caused by the passing light beam are visible on each processed surface. Those lines give raise to the bevelled surface roughness and make it wrinkled, as presented in figure 3. The wrinkles are especially intensive close to the top surface of the sample, where laser beam remains well collimated.

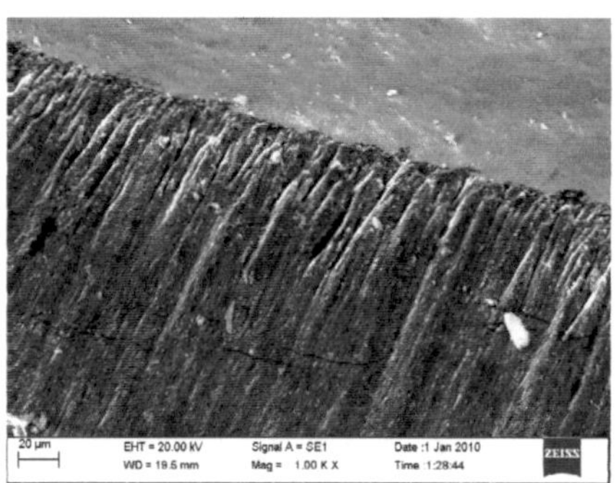

Figure 3. SEM image of laser bevelled surface (30°).

During the laser machining, locally heated SiC reacts with oxygen from the atmosphere leading to formation of CO_2 (volatile) and SiO_2 (solid), which can be seen in figure 2c. An accumulation of silicon oxide grains may interfere with SiC laser processing making it less controllable and inaccurate.

3. Conclusions

In the paper the new approach for bevel termination fabrication for SiC high-voltage devices is presented. The green laser equipped with beam scanner and dedicated rotating sample holder has been employed to verify the results of silicon carbide laser machining. The structures obtained were characterised using electron microscopy. Observed results proved that it is possible to create a bevelled surface at the edges of the round silicon carbide device and therefore the presented method may be useful for termination of the device's junction. At present the process remains immature and results with rough and uneven surface, but since the surface quality is strongly dependant on the laser beam and sample holder parameters many ways of improvement are possible and will be examined.

4. References

[1] Baliga B J 2008 *Fundamentals of Power Semiconductor Devices* (Springer)
[2] Huang X, Baliga B J 2012 *IEEE Electron Device Letters* **33(11)** 1592-1594
[3] Biscarrat J, Michaud J F, Collard E, Alquier D 2013 *Materials Science Forum* **740-742** 825-828
[4] Cheng J, Liu C S, Shang S, Liu D, Perrie W, Dearden G, Watkins K 2013 *Optics and Laser Technology* **46(1)** 88-102
[5] Dahotre N, Samant A 2011 *Laser Machining of Advanced Materials* (CRC Press)
[6] Vanko G, Hudek P, Zehetner J, Dzuba J, Choleva P, Kutis V, Vallo M, Ryger I, Lalinsky T 2013 *Microelectronic Engineering* **110** 260-264
[7] Pangovski K, Sparkes M, Cockburn A, O'Neill W, The P S, Lin D, Richardson D 2014 *IEEE Journal on Selected Topics in Quantum Electronics* **20(5)** 6742663

Journal of Physics: Conference Series **709** (2016) 012006 doi:10.1088/1742-6596/709/1/012006

Porous silicon Bragg reflectors on multi-crystalline silicon wafer with p-n junction

I I Ivanov[1,3], V A Skryshevsky[1], O S Kyslovets[1], T Nychyporuk[2] and M Lemiti[2]

[1]Institute of High Technologies, Taras Shevchenko National University of Kyiv, 64 Volodymyrska street, Kyiv, Ukraine, 01601,

[2]Institut des Nanotechnologies de Lyon, UMR CNRS 5270, Université de Lyon, INSA Lyon, Bct. Blaise Pascal, 7 Avenue Jean Capelle, 69621 Villeurbanne Cedex, France

[3]E-mail: ivancko@gmail.com

Abstract. Bragg reflectors consisting of the sequence of dielectric layers are considered to create p-n junction solar cells (SC) with improved efficiency in the longwave spectral range. Bragg mirrors (BM) based on porous silicon (PS) mutilayers at the backside of single crystalline and multicrystalline silicon wafer were formed by electrochemically etching. Maximal experimental reflectivity for BM on multicrystalline substrate achieves 62% due to the natural crystallites disorientation of multicrystalline substrate, whereas for single crystalline silicon the reflectivity in maximum is 87%. BM was formed also on rear side of multicrystalline silicon wafer with p-n junction.

1. Introduction

The increasing efficiency of SC and the reducing their cost is the main direction of development of solar energy through direct conversion of solar energy into electricity. The concept of third generation solar cells declares the need to improve the efficiency of existing SC using new technology opportunities and the creation of SC that operate on new principles [1]. Process parameters and production phases optimization of existing designs of silicon SC is realized to implement this concept. SC with quantum wells [2] and quantum dots [3], organic material [4] and dyes based dispersed SC [5], thin film SC [6] are actively investigated.

There are some main power losses mechanisms for single p-n junction SC [7]. The first one is due to thermalisation losses of photogenerated charge carriers. The second one is inability of SC to absorb the photons with energies less than the semiconductor band gap. The cost of commercial Si SC can be decreased at the using of thinner wafer. However, it results in the reduction of absorption of low energy photons. One of the way to overcome this problem is the modification of SC design with the improved schema of photons trapping having absorption length $L{\cdot}\alpha=1/\alpha>L_w$ (figure 1), where α is absorption coefficient of semiconductor wafer, L_w is wafer thickness.

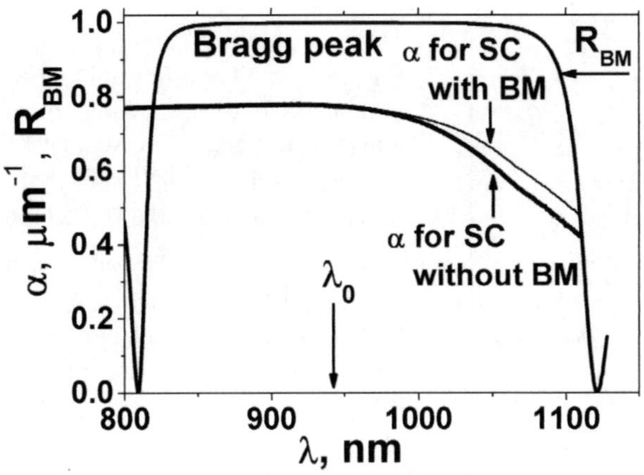

Figure1. Calculated spectral dependence of reflection coefficient of BM, spectral dependence of absorption coefficient of silicon SC with and without rear BM.

Pyramids texturing [8] and antireflection layers [9] are using to reduce reflection from front SC. Photon traps [10], rear reflectors [11] and texturing [8] are using to increase photon effective absorption length due multiple reflections of photons inside wafer.

PS as Si-technology compatible material can be used to form antireflection coating and rear reflector for Si SC [12]. PS is grown by electrochemical etching from crystalline Si in electrolytes containing hydrofluoric acid (HF) [13]. PS layer parameters are depended on etching current density, etching time, solution composition and wafer crystal-lattice orientation. PS rear reflector are used as base for further deposition of thin film SC by epitaxy or chemical vapor deposition [14]. Modern commercial SC are fabricated on the base of single crystalline as well as multicrystalline Si wafers, so spectral dependences of reflectivity for single and multicrystalline Si based reflectors have to be differ due to porosity dependence on crystal-lattice orientations [15].

In present work we consider structures with rear PS Bragg mirrors on single crystalline and multicrystalline silicon substrates and propose the method to fabricate PS Bragg mirror on two-side textured multicrystalline silicon wafer with formed p-n junction.

2. Experiment

Bragg mirror can be used, as rear reflector for multi-passing of IR photons within the SC. BM has high reflectivity within certain spectral range (figure 1). BM is structure formed from multiple layers of alternating materials with different refraction indexes. In simple case BM is the stack of two layers with optical thicknesses $n_H \cdot l_H = n_L \cdot l_L = \lambda_0/4$, where λ_0 is the wavelength of BM maximum of reflection, n_H, n_L are refraction indexes of layers, l_H, l_L are thicknesses of layers [16].

Bragg mirrors were fabricated by electrochemical etching of single- and multi-crystalline p-type silicon wafer. Electrochemical treatments were in HF/C_2H_5OH solution mixed in the ratio 4:1. Diameter of all samples is 2.5 cm, area is 4.9 cm^2. "Step-like" current treatment profile with sharp current values changes was used.

All measurements of reflection coefficients were performed using standard reflectivity measurement system consisting of a light source, computer controlled monochromator, chopper, Si-photodiode as light sensor, selective multimeter, ADC board. Measurements were performed in dark room in isolated black box. Reflection coefficients were measured using integrating sphere to take into account diffuse component of reflected light. Certified samples with known spectral reflectivity dependence were used for measurement system calibration before every measurement of reflection coefficient.

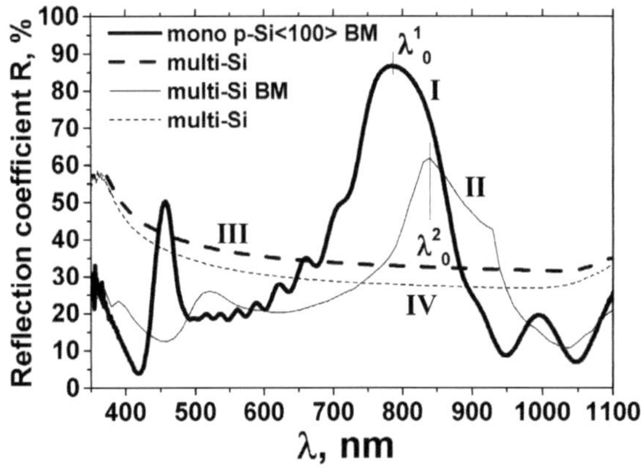

Figure 2. Experimental spectral dependences of reflection of BM grown on single crystalline p-Si (<100> orientation) both-side polished wafer λ_0^1=788 nm (curve I), BM grown on multicrystalline p-Si wafer λ_0^2=840 nm (curve II), single crystalline p-Si (<100> orientation) both-side polished wafer (curve III), multicrystalline p-Si wafer (curve IV).

Reflection coefficient of BM on 2-side polished single crystalline silicon wafer are $R^1_{max}(\lambda_0^1 = 788\text{ nm})$=86.5% (figure 2, curve I). Etching parameters and obtained refraction indexes were the following: j_H=1.8 mA/cm^2, n_H=2.1; j_L=623 mA/cm^2, n_L=1.4. Reflection coefficient of BM on 2-side polished multicrystalline silicon wafer are $R^2_{max}(\lambda_0^2 = 840\text{ nm})$=61.8% (figure 2, curve II).

In the vicinity of Bragg peak (BP) the reflection coefficients of BMs fabricated on mono- and multi-Si is larger than reflection coefficients of mono- and multi-Si wafer (figure 2, curve III and IV respectively). Multicrystalline silicon wafer consists of elementary clusters with different crystal-lattice orientation. So, due to dependence of growth rate of PS layers on crystal lattice orientation, effective total reflection coefficient of such BM is defined as weighted sum of reflection coefficients of separate elementary clusters as well as clusters' area.

To fabricate BM on multicrystalline silicon wafer such etching parameters as grown rate $\upsilon_{PS}(j)$ and refraction index n(j) were taken for crystal lattice orientation <100>. Crystal lattice orientation within elemental cluster and area of cluster varies from wafer to wafer and, therefore, total reflection coefficient of multicrystalline based BM is less than for single crystalline based BM: $R^2_{max} < R^1_{max}$. Such BM can be used as base to fabricate SC with rear reflector. SC can be formed on the top of BM using such technologies as chemical vapor deposition, molecular-beam epitaxy.

BM formed on both-side textured structure with formed p-n junction was formed. Wafer parameters are the following: wafer thickness 200 μm (N_a=10^{16} cm^{-3}), phosphorus doped (N_d=10^{20}cm^{-3}) emitter thickness is 300 nm, sample area is 4.9 cm^2. Alternative current profile was used for electrochemical treatment (figure3, insert). BM has to be formed on external base interface, so structure with p-n junction is under reverse bias condition for traditional "step-like" current treatment profile and, therefore, alternative current profile was used to grow BM. Alternative current treatment frequency was 100 Hz. Treatment amplitudes I_L and I_H (j_H=10 mA/cm^2, j_L=348 mA/cm^2) were chosen to obtain required PS layers growth rates and refraction indexes (n_H=1.9, n_L=1.6). Treatment parameters j_H, $j_L t_H$, t_L were chosen to obtain BP maximum at λ_0^3=570 nm. Measured reflection coefficient is $R^3_{max}(\lambda_0^3 = 572\text{ nm})$=32.4% (figure 3).

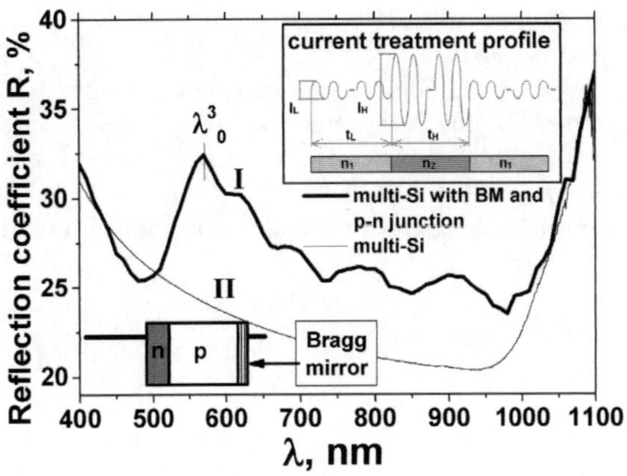

Figure 3. Spectral dependences of reflection of BM which was grown on multicrystalline both-side textured p-Si wafer with formed p-n junction λ_0^3=572 nm (curve I), multicrystalline both-side textured p-Si wafer (curve II).

BP shape (figure 3, curve I) is worse comparatively to shape of BP of Bragg mirrors (figure 2) formed on single- and multicrystalline both-side polished Si wafer. It can be explained by rear contact interface roughness due texturing. Wafer with preliminarily formed texture has not flat contact with electrode during electrochemical treatment, non-uniform PS layer growth is observed. The spectral dependence of reflectivity before PS forming is shown in figure 3 (curve II). The curve of reflection coefficient for structure with BM lies above than for structure without BM. Electrochemical treatment of textured wafer with existing p-n junction is complicated process and requires additional optimization.

The elaborated structures with porous silicon BM show high stability and reversibility. Particularly, the 30 min annealing at 900°C did not change the optical parameters of structures.

3. Conclusion

Bragg reflector on the base of porous silicon can increase absorption of IR photons due to multi-passing IR photons and properly absorption length increasing inside monocrystalline and multicrystalline Si wafer. Electrochemical treatment with sharp "step-like" current treatment profile can be used to fabricate BM with maximum measured reflectivity 86% for monocrystalline Si wafer and 61% for multicrystalline Si wafer. Total reflectivity of multicrystalline Si wafer is defined by weighted sum of reflection coefficients of separate elementary wafer clusters. BM can be formed on structure with formed p-n junction using alternative current treatment profile.

References

[1] Green M A 2003*Third Generation Photovoltaics: Advanced Solar Energy Conversion* (Berlin: Springer)
[2] Mazzer M, Barnham K, Ballard I M, Bessiere A, Ioannides A, Johnson D C, Lynch M C, Tibbits T N D, Roberts J S, Hill G, Calder C 2006 Progress in quantum well solar cells *Thin Solid Films* **511-512** 76-83
[3] Jun H, Careem M and Arof A 2013 *Renew. Sustainable Energy Rev.* **22** 148-167
[4] Matthew W 2012 *Sol. Energy Mater. Sol. Cells.* **107** 87-111
[5] Gonga J, Liang J, and Sumathya K 2012 *Renew. Sustainable Energy Rev.***16** 5848–5860
[6] Green M A 2007 *J. Mater. Sci. - Mater. Electron.* **18** 15-19
[7] Wolf M and Rauschenbach H. **1976** *Solar Cells* (New-York : IEEE Press)
[8] Papet P, Nichiporuk O, Kaminski A and Lemiti M 2006 *Sol. Energy Mater. Sol. Cells* **90** 2319-2328
[9] Sahouane N and Zerga A 2014 *Energy Procedia* **44** 118–125
[10] Christoph J, Goldschmidt B, Kirchartzc T and Bläsi B 2009 *Sol. Energy Mater. Sol. Cells* **93** 1721-1727

[11] Cui H, Patrick R and Green M A 2006 *Energy Procedia* **90** 2319–2328
[12] Aziz W, Ramizy A, Ibrahim K, Hassan Z and Omar K 2011 *Optik* **122** 1462–1465
[13] Ivanov I, Skryshevsky V, Serdiuk T and Lysenko V 2012 *Sens. Actuators B* **174** 521-526
[14] Duerinckx F, Kuzma-Filipek I, Nieuwenhuysen K and Beaucarne G. 2006 *IEEE Electron Device Lett.* **27** 837−839
[15] Liua J, Liua B, Shena Z and Liua J. 2012 *Sol. Energy* **86** 3004–3008
[16] Ivanov I, Skryshevsky V, Nychyporuk T, Lemiti M, Makarov A, Klyui N and Tretyak O 2013 *Renew. Energy* **55**, 79-84

Measurements of thermal parameters of solar modules

K Górecki and E Krac

Gdynia Maritime University, Department of Marine Electronics, Morska 83, Gdynia, Poland

E-mail: k.gorecki@we.am.gdynia.pl

Abstract. In the paper the methods of measuring thermal parameters of photovoltaic panels - transient thermal impedance and the absorption factor of light-radiation are presented. The manner of realising these methods is described and the results of measurements of the considered thermal parameters of selected photovoltaic panels are presented. The influence of such selected factors as a type of the investigated panel and its mounting manner on transient thermal impedance of the considered panels is also discussed.

1. Introduction

Solar cells are more and more often used as a source of electrical energy [1]. Characteristics of these semiconductor devices strongly depend on temperature [2, 3], which also in a decisive manner influences reliability of electronic equipment [4, 5]. The internal temperature of the considered devices is higher than the ambient temperature as a result of thermal phenomena - self-heating and the energy exchange of solar radiation into heat [3, 6, 7].

In order to take into account thermal phenomena in computer analyses, electrothermal models of electronic elements are indispensable. The electrothermal model of solar cells was proposed in the paper [3]. In this model self-heating phenomenon is characterized by transient thermal impedance $Z_{th}(t)$, which is typically represented by the following dependence [6, 8]

$$Z_{th}(t) = R_{th} \cdot \left[1 - \sum_{i=1}^{N} a_i \cdot \exp\left(-\frac{t}{\tau_{thi}}\right)\right] \tag{1}$$

where R_{th} denotes thermal resistance, N – number of thermal time constants τ_{thi} connected with coefficients a_i.

Methods of measuring thermal resistance of solar cells are presented, among others, in the paper [9]. However, in the real photovoltaic systems, the panels containing many solar cells connected with one another electrically and installed on the common basis are used [7]. But in literature there is no description of methods of thermal parameters measurements of such panels.

In this paper the method of measuring transient thermal impedance of photovoltaic panels and the method of estimating the absorption factor of radiation by such a panel are presented. Theoretical considerations are illustrated with the results of measurements.

2. Measurements method

In accordance with the well-known from the papers [10, 11] idea of measuring transient thermal impedance of semiconductor devices, the value of this parameter can be measured using the so called cooling curve. According to this idea, power of the well-known value P should be dissipated in the examined device until the steady state is obtained, and then after switching-off this power, the waveform of the internal temperature of the examined element $T_j(t)$ should be measured. In the last step of the measurement the values of $Z_{th}(t)$ are calculated using the following formula

$$Z_{th}(t) = \frac{T_j(t=0) - T_j(t)}{P} \tag{2}$$

where $t = 0$ denotes the moment of switching-off the power.

In the case of the photovoltaic panel one ought to assure the conditions of the $Z_{th}(t)$ measurements, in which the only reason for the change of the internal temperature of the panel is self-heating. Therefore, the measurement is performed for not lighted up panel situated in the light-tight chamber. The temperature of the panel T_j is measured by means of the thermo-hunter connected with the computer, which registers the values of this temperature in the selected points of time. The power P is equal to the product of the voltage and the current of this panel measured before switching-off this power.

In turn, the absorption factor of radiation a is measured at the steady-state for the panel lighted up using the following formula.

$$a = \frac{1}{S \cdot p_e} \cdot \left(\frac{T_j - T_a}{R_{th}} - U_{nom} \cdot I \right) \tag{3}$$

where I means the current of the panel, U_{nom} - voltage at the opening output of the panel, S - the active area of the panel, while p_e is power density of radiation on the surface of the panel.

Now then, for the purpose of measuring the value of the absorption factor of radiation, one should first measure thermal resistance of the panel R_{th}, and then read the voltage U_{nom} for the lighted up panel at the zero-current; next, one should measure the value of temperature of the lighted up panel by means of the thermo-hunter at the current of the panel I.

The measuring-methods described above are realized in the measuring-system shown in Figure 1.

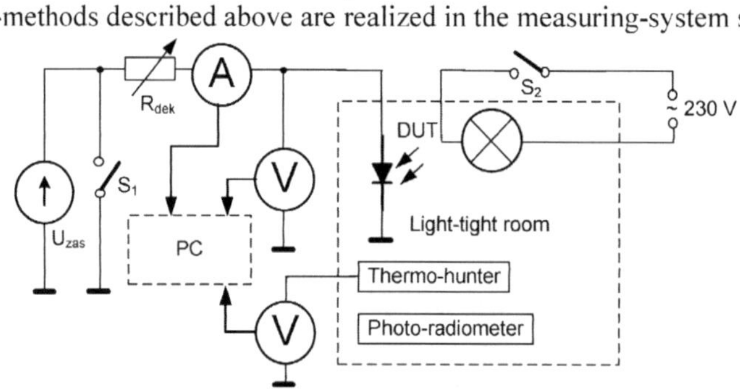

Figure 1. The measurement-system of transient thermal impedance and the absorbtion factor of radiation of the photovoltaic panel

During the measurement of transient thermal impedance the switch S_2 is opened, and the voltage source U_{zas} together with the resistor R_{dek} make possible the regulation of the supply current of the examined panel DUT. In the moment $t = 0$ the switch S_1 is opened, making possible the flow of the current through the examined panel. The temperature of the panel is measured by the thermo-hunter and registered by the PC.

In turn, during the measurement of the absorption factor the switches S_1 and S_2 are closed. The light source lights up the examined panel, power density of radiation is measured at the steady-state by means of the photo-radiometer, and temperature of the panel is measured by means of the thermo-hunter.

3. Results of measurements

The measuring methods presented in the second section were used to measure thermal parameters of the selected photovoltaic panels operating at different conditions of the power supply and cooling.

In Figure 2 waveforms of transient thermal impedance of the panel MK-240 containing 60 poly-crystalline silicon solar cells connected serially are presented. This panel has the dimensions 1000 x 1681 x 50 mm, and the nominal power dissipated in this panel amounts to about 200 W. In this figure, the curve a means transient thermal impedance for the panel situated horizontally on the steel-rack, the curve b - for the panel situated on the same rack slanting at a little angle of 30° to the horizon, the curve c - for the panel situated on the same rack slanting at a little angle of 60° to the horizon, curve d - for the panel situated horizontally on the rack and separated from it with the wooden plate, while the curve e - for the panel pendant on 4 metal-brackets. The measurements were performed at dissipated power equalled to about 310 W.

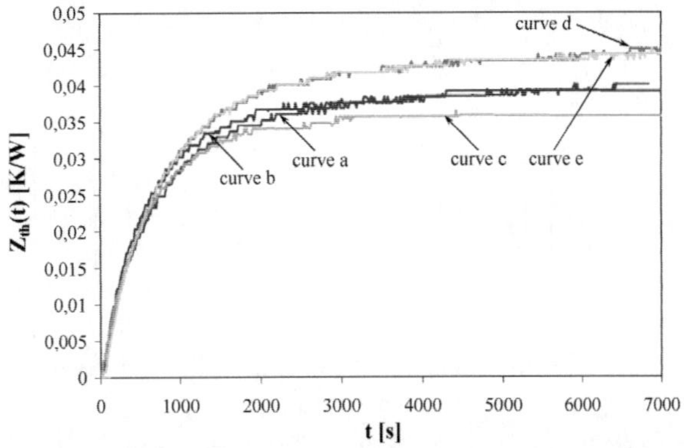

Figure 2. Measured waveforms of transient thermal impedance of the panel MK-240 operating at different cooling conditions

As it is visible in Figure 2, time indispensable to obtain the steady state and the value of thermal resistance depends on the operating conditions of the panel. The steady state is obtained after 3000 – 5000 s after switching on the power supply of the panel. The least value of thermal resistance, equal to 0.0365 K/W, is obtained for the panel situated on the stand metal and slanting at an angle of 60° to the horizon, the biggest - equal to 0.045 K/W - for the panel arranged horizontally on the metal stand - and separated from it by means of the 4 mm wooden plate (of the heat conductivity coefficient equal to 0.16 W/m/K). It is worth noticing that together with an increase of the depression angle of the panel with relation to the horizon in the range from 0 to 60°, one observes a decrease of thermal resistance by even about 20%. It is also easy to notice that the mounting base, to which the panel is mounted has the essential meaning in the process of heat removal from the panel. In the case of the direct contact of the panel with this base one obtains values of thermal resistance by even about 10% lower than in the situation, when between the panel and the metal-base a material weakly conducting heat is inserted.

Dynamic properties of the process of heat removal dissipated in the panel can be characterized by spectrum of thermal time constants [8]. The spectrum of thermal time constants for the panel MK-240 are shown in Figure 3. In this figure the colours used to present this spectrum are the same as corresponding to them waveforms of transient thermal impedances denoted as curves a-e in Figure 2. Values of parameters a_i and τ_{thi}, existing in the equation (1), were marked with the use of the algorithm described in the paper [12]. This algorithm uses measured waveforms of $Z_{th}(t)$.

As it can be observed, at all the considered cooling conditions at least two thermal time constants appear. Thermal time constants of values in the range from 200 to 600 s have the prevailing meaning (the corresponding to them values of parameters a_i have the biggest value). The longest thermal time constants in the range from 800 to 2300 s decide about time indispensible to obtain the steady state. Comparing the curves shown in Figure 2 and Figure 3 it can be noticed that to worse cooling conditions correspond both higher values of thermal resistance and higher values of thermal time constants.

Except the panel MK-240 the measurements of transient thermal impedance were performed also for other photovoltaic panels: the amorphous panel SG-HN-120-GG of the dimensions 1400 x 1100 x 7.1 mm and the nominal output power equal to 120 W, the monocrystalline panel SGM -250D of the dimensions 1650 x 990 x 40 mm and the nominal output power equal to 250 W and the poly-crystalline panel Q.PRO-G3 250-265 of the dimensions 1670 x 1000 x 35 mm and the nominal output power equalled to 250 W.

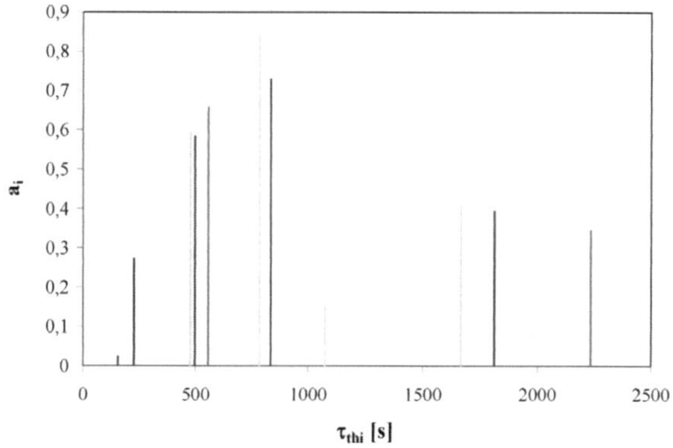

Figure 3. Spectrum of thermal time constants of the panel MK-240 operating at different cooling conditions

Figure 4 presents the measured waveforms of transient thermal impedance of the considered panels at worse cooling conditions (the panel situated horizontally on the rack and separated from it with the wooden plate), whereas Figure 5 - at the optimum cooling conditions (the panel situated on the same rack slanting at a little angle of 60° to the horizon). In these figures curves a correspond to the panel MK-240, curves b - the panel SG-HN-120-GG, curves c - the panel SGM-250D, whereas curves d - the panel Q.PRO-G3 250-265.

As one can notice, higher differences between the waveforms of transient thermal impedance are obtained at worse cooling conditions. For all the considered panels time indispensable to obtain the steady state equals 2 hours. Values of thermal resistance obtained for panels SG-HN-120-GG and Q.PRO-G3 250-265 differ by over 25%. The panel SG-HN-120-GG differs constructively from other panels, ie. it does not possess the aluminium-frame, but it is situated on the glass basis.

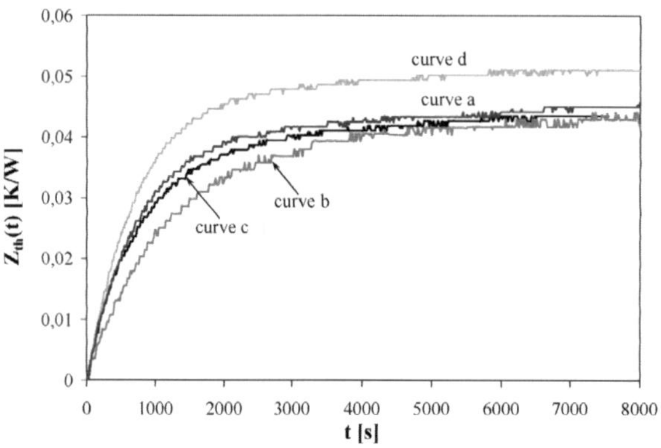

Figure 4. Measured waveforms of transient thermal impedance of the selected panels situated horizontally on the rack and separated from it with the wooden plate

In turn, the visible in Figure 5 waveforms $Z_{th}(t)$ prove that the use of the metal rack bracket under the large angle to the horizon assures essential improvement in the cooling efficiency of the considered panels. It is worth noticing that in the considered case differences are visible in heat capacitances of each panel appearing at different times indispensible to obtain the steady state. Simultaneously one observes small differences in the value of thermal resistance of each panel. These differences exceed 15%. The time indispensable to obtain the thermally steady state in the investigated panels reaches even 6000 s.

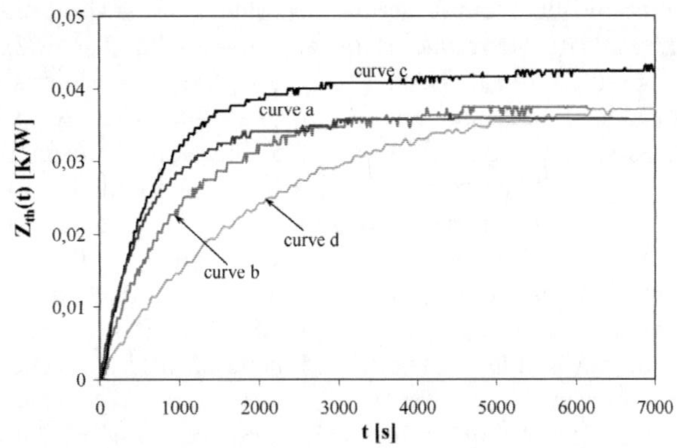

Figure 5. Measured waveforms of transient thermal impedance of the selected panels situated on the rack slanting at a little angle of 60° to the horizon

In Figure 6 the spectrum thermal time constants corresponding to the waveforms of transient thermal impedances shown in Figure 5 are presented. It is easy to observe large divergences in values of thermal time constants characterizing each of the considered panels. Values of the longest thermal time constants, decisive about time indispensable to obtain the thermally steady state of the investigated panels, accept values from the range from 1000 s for the panel MK-240 to 5000 s for the monocrystalline panel SGM -250D. The transient thermal impedance of the amorphous panel, as the only one of the investigated panels, is described by means of only one thermal time constant. Thermal proprieties of all the remaining panels are characterized by two thermal time constants.

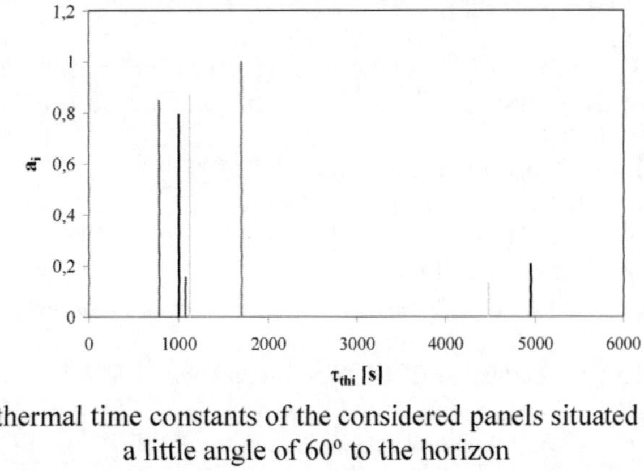

Figure 6. Spectrum of thermal time constants of the considered panels situated on the rack slanting at a little angle of 60° to the horizon

Using the method described in the second section, the absorption factor of radiation is measured for the selected photovoltaic panels. During the measurements power density of radiation on the surface of the examined panel amounted to about 175 W/m^2. The load current of the panel amounted to about 3 mA. The panel was pendant on 4 metal-brackets. Time indispensable to obtain the steady state exceeded 2 hours. In accordance with the equation (3) the values of the absorption factor of radiation are obtained and they equal 2.99 for the polycrystalline panel MK-240 and 2.405 for the monocrystalline panel SGM-250D. The obtained values of the considered parameter mean that the absorption of solar radiation can cause the excess of the panel internal temperature over the ambient temperature equal to even 200 K at the maximum allowable value of the power density of solar radiation (1000 W/m^2).

4. Conclusions

In the paper the methods of measuring thermal parameters (transient thermal impedance and absorption factor of radiation) of photovoltaic panels are presented and some results of measurements of these parameters for the selected panels operating at different conditions of their mounting are shown. The presented results of measurements confirm usefulness of the worked out measuring-methods. The investigations were performed by means of the worked out methods. The influence of the manner of mounting photovoltaic panels on their transient thermal impedance confirmed that among others the attack angle of the panel in relation to the horizon and thermal conductance of material situated between the panel and the mounting-base had the essential meaning.

The obtained waveforms of transient thermal impedance of the panels and the measured values of the absorption factor of radiation show that for the nominal value of the output power of the examined panels and for the power of sun lighting observed in Poland an increase in temperature of the panel above the ambient temperature can exceed tens kelvins. Such a large increase of temperature can cause essential depreciation of the output voltage and the output power [3]. Therefore, optimization of cooling conditions of photovoltaic panels can cause an increase in their efficiency.

5. References

[1] Mulvanet D 2014 *IEEE Spectrum* **51** (9) 26-29
[2] Castaner L, Silvestre S 2002 *Modelling photovoltaic systems using Pspice* (John Wiley&Sons)
[3] Górecki K, Górecki P, Paduch K 2014 *Journal of Physics: Conference Series* **494** 012007
[4] Castellazzi A, Gerstenmaier YC, Kraus R, Wachutka GKM 2006 *IEEE Transactions on Power Electronics* **21** (3) 603-612
[5] Reynolds FH 1982 Measuring and modeling integrated circuit failure rates *Eurocon'82, Copenhagen: Reliability in Electrical and Electronic Components and Systems. North Holland* **1** 36-45
[6] Bagnoli PE, Casarosa C, Ciampi M, Dallago E 1998 *IEEE Transactions on Power Electronics* **13** (6) 1208-1219
[7] Górecki K, Krac E, Zarębski J 2015 Photo-electro-thermal characteristics of photovoltaic panels *Springer Proceedings in Energy, 2nd International Congress on energy Efficiency and Energy Related Materials ENEFM 2014* Oludeniz 45-51
[8] Szekely V 1997 *Microelectronic Journal* **28** (3) 277-292
[9] Krac E, Górecki K 2014 *Zeszyty Naukowe AMG* **84** 56-65
[10] Blackburn DL, Oettinger FF 1976 *IEEE Transactions on Industrial Electronics and Control Instrumentation* **IECI-22** (2) 134-141
[11] Zarębski J, Górecki K 2007 *IEEE Transactions on Components and Packaging Technologies* **30** (4) 627 – 631
[12] Górecki K, Rogalska M, Zarębski J 2014 *Microelectronics Reliability* **54** (5) 978-984

Investigations of DC power supplies with optoelectronic transducers and RF energy converters

B Guzowski[1], **R Gozdur, L Bernacki, M Lakomski**

Department of Semiconductor and Optoelectronic Devices, Lodz University of Technology, Wolczanska 211/215, 90-924 Lodz, Poland

[1]E-mail: bartlomiej.guzowski@p.lodz.pl

Abstract. Fiber Distribution Cabinets (FDC) monitoring systems are increasingly popular. However it is difficult to realize such system in passive FDC, due to lack of source of power supply. In this paper investigation of four different DC power supplies with optoelectronic transducers is described. Two converters: photovoltaic power converter and PIN photodiode can convert the light transmitted through the optical fiber to electric energy. Solar cell and antenna RF-PCB are also tested. Results presented in this paper clearly demonstrate that it is possible to build monitoring system in passive FDC. During the tests maximum obtained output power was 11 mW. However all converters provided enough power to excite 32-bit microcontroller with ARM-cores and digital thermometer.

1. Introduction

Fiber optic technology is used in ever-increasing applications due to its inherent advantages (lower weight, EMI/RFI immunity, higher bandwidths and distances) over copper. Billions of packets of digital data wend their way around the world, through switches, long haul optical fibers, metro rings, and assorted other access topologies. At each node incoming fibers are patched from Fiber Distribution Cabinets (FDC) to equipment and back again. Hundreds of fibers can be installed in cables taking relatively little space. However, terminating these fibers with connectors increases the required space in Optical Distribution Frames (ODF).

Large Data Centers have also a hundreds of thousands of optical fiber cables and connections, and there is a need to track the connectivity of these trunk and jumper cables to assure proper network interconnection among the servers, switches, routers, and storage equipment. Therefore, it is extremely hard to track and maintain the topology of the network with very high accuracy given the large number of manually changeable interconnection points in the network.

Ultra – low power monitoring systems are one of solutions that can improve the efficiency and accuracy of maintaining the network connectivity. Hence these systems can be supply by optoelectronic transducers, it is possible to integrate monitoring system with passive FDC.

In this paper detailed analysis of four different DC power supplies with optoelectronic and RF transducers is described. The most suitable option is to convert light transmitted through the opitcal fiber to electric power. It is stabile power source and it can be used regardless of weather or environment in which the FDC is placed. For that reason the photovoltaic power converter PPC – 6E by JDSU [1] and a silicon PIN photodiode BPW24R from Vishay are tested. PPC – 6E is designed to convert 790 – 850 nm light into electrical power up to 6 V, with output power up to 500 mW. Assuming temporary or permanent illumination in Data Centers or server rooms it is possible to convert the light using solar cells. Therefore popular polycrystalline silicon solar cells are also tested. Due to the fact, that more and more devices broadcast the RF energy the ability to harvest RF energy, from ambient or dedicated sources is more likely. Hence, the last type of investigated converters was a directional antenna RF – PCB designed for GSM band 880 – 915 MHz.

2. Electromagnetic radiation energy harvesting devices

Monitoring system requires to operate electrical power. Commercially available ultra – low power microchips or sensor systems need less than 1 mW power [2, 3]. Since majority of FDC are totally passive it is necessary to introduce to cabinets some kind of power supply. Although the non-renewable energy sources, such as batteries can be used, these sources do not provide long life operating or autonomy [4]. A solution that can enable to do this is using energy – harvesting source. Great number of sources of energy and conversion devices could been considered for energy harvesting [5, 6] and in order to compare different approaches, it is useful to consider the criteria for comparison. However implementation of these system in passive FDC results in the power supplies acting on the basis of electromagnetic radiation seem to be the best option. Electromagnetic radiation converters are, either in their form of RF converters or solar cells operating in VIS or IR light. In RF converters high power electromagnetic energy from nearby source is directed to the device and converted to the voltage. However the RF converters are very strongly dependent on environment, architecture of a building and require a transmitter in a close range. In table 1 output voltage of nowadays RF converters are shown.

Table 1. Output voltages of RF converters.

Output voltage	Wavelength	Realization
3.22 V @ 3 kΩ load	980 MHz	Arrawatia [7]
3.00 V @ 100 kΩ load	945 MHz	Saraiva [8]
0.68 V @ 50 kΩ load	1.8 GHz	Arrawatia [9]
2.05 V @ 100 kΩ load	900 MHz	You [10]

Solar cell convert the energy of light directly into electricity. This mature technology has wide range of applications. Efficiency of outdoor solar systems depend on e.g. weather conditions or location. In indoor installation where the irradiance is around 3.5 – 20 W/m^2 [6] output power depends mostly on efficiency of solar cell. Since the efficiency of the best of solar cells is around 25 – 30% [11-13] it is possible to obtain 30 – 50 mW from 1 cm^2 [14], if the solar irradiation is 1000 W/m^2. However, due to the fact, that FDC and ODF are often placed in rooms without permanent illumination, or with the artificial illumination usually with low power density using a solar cells seems to be limited. Therefore photovoltaic power converters (PPC) dedicated to converting the light coming out from optical fibers have to be also considered. As mentioned before, those converters can operate in almost all environment conditions and constructions. These millimeter size structures are typically made of GaAs and dedicated to the wavelength range of 790 nm to 850 nm [15]. Illuminating PPC with 400 mW of optical power (50 W/cm^2), up to 133 mW electrical power can be harvested [16]. However it must be taken into account that one or more ports in ODF will need to be separated to transfer power from light source.

3. Supply systems configurations

In order to develop a supply system using alternative energy sources, a number of experimental tests on the selected commercially available transducers, and RF/DC and DC/DC converter. Summary of the test transducer which will be used as the energy sources is shown in table 2.

Table 2. Tested converters.

Type of converter	Name	Size
Photovoltaic power converter 790 – 850 nm	PPC – 6E, from JDSU	Ø 2.39 mm
Si Photodiode max. @ 900 nm	BPW24R, from Vishay	0,78 mm^2
Polycrystalline Si solar cells	AM – 5412CAR, from Panasonic	50 x 33 mm
Directional antenna RF-PCB on the EGSM band 880 – 915 MHz	H 122°/V 68°, gain 6,1 dBi, from Powercast	175 x 50 mm

Figure 1 shows a set of optimized RF/DC and DC/DC converter systems, designed for elements listed in table 3. Using different types of converters results from the fact that each of the tested elements have different input and output parameters e.g. the range of the input/output voltage, output resistance and the limit of continuous output current.

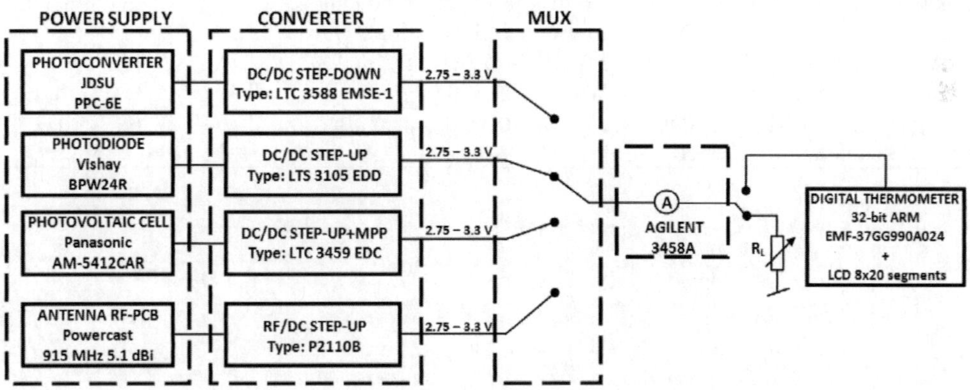

Figure 1. A block diagram of the measuring system.

Each type of converter provides DC power supply at an output voltage ranging from 2.75 V to 3.3 V and load specified by the type of tested converter. The decade resistor was used as primary test load circuit. In addition, a digital thermometer, built on the 32-bit ARM Cortex-M3 microcontroller with LCD alphanumeric display was used as test load as well.

4. Determination the power output of power supply systems

The first stage of the research was to determine maximum power transfer from supply circuits. For this purpose we used an adjustable resistive load and direct current measurement system using Agilent 3458 multimeter. In each of the cases load resistance R_L was regulated in order to determine the best conditions of matching the load to the source. As a result we determined threshold value of load resistance R_L at which we observed 10% drop in the value of output voltage of power supply. In addition we determined time period t_{on} from the moment when we started excite the transducer until the change of logic state of PGOOD each converter. Later in the research we used as a load a real MCU system, in which the microcontroller made periodical temperature measurements in deep sleep mode. In this configuration, the system nominally charged to the power supply current of 1.1 μA.

Methods excitation of tested sensors differ depending on their type. Photovoltaic cell was excited by visible LED light with illuminance of 300 lux. The receiving system equipped with a directional antenna (H 122°/ V 68°) at 880 – 915 MHz GSM band was placed at 60 cm distance from the 3 W EGSM transmitter. PPC-6E and Si photodiode were excited by a powerful IR LED array SFH4750 from OSRAM and laser source 850 nm from FIS Deluxe Mini OTDR. Optical power level was set to 3 W for IR LED and 0.6 mW for laser source. Light coming out from laser source was distributed to converters through 1 m length multi-mode optical fiber (MMF) OM4 class. Optical fiber was connected to converters using ST optical fiber connectors.

The distance between the IR LED and Si photodiode and PPC – 6E was always individual set to obtain three constant and repetitive values of illumination excitation. The results of measurements of the load current I_L at the threshold value of R_L, max. output power of power supply and t_{on} time are shown in table 3.

Table 3. Results of measurements.

Converter	Energy source	I_L [mA]	P_m [mW]	R_L [kΩ]	t_{on} [s]
Photovoltaic power converter 790 – 850 nm	0.6 mW laser source 850nm from FIS Deluxe Mini OTDR	0.01	0.04	380	30(
Photovoltaic power converter 790 – 850 nm	3W SFH4750 from OSRAM	0.82	2.72	4.0	3
Si Photodiode max.@ 900nm	0.6 mW laser source 850nm from FIS Deluxe Mini OTDR	-	-	-	-
Si Photodiode max.@ 900nm	3W SFH4750 from OSRAM	2.75	9.08	1.2	3
Polycrystalline Si solar cells	White LED 300 lx	0.41	1.36	8.0	5
Directional antenna RF – PCB on the EGSM band 880 – 915 MHz	3W EGSM 900 Transmitter 0.6 m	2.36	7.78	1.4	3

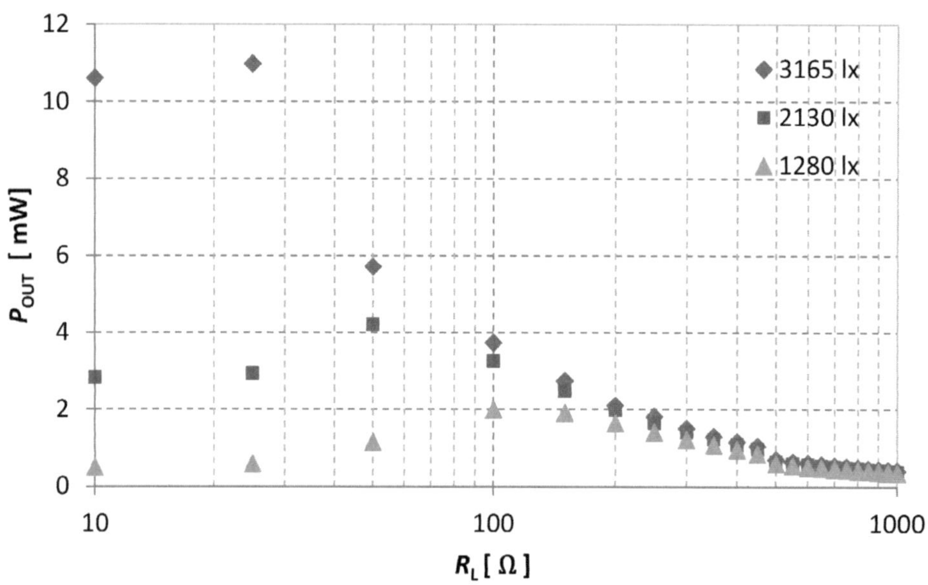

Figure 2. Influence of matching load R_L on the output power P_{OUT} of silicon PIN photodiode BPW24R for different illumination excitation.

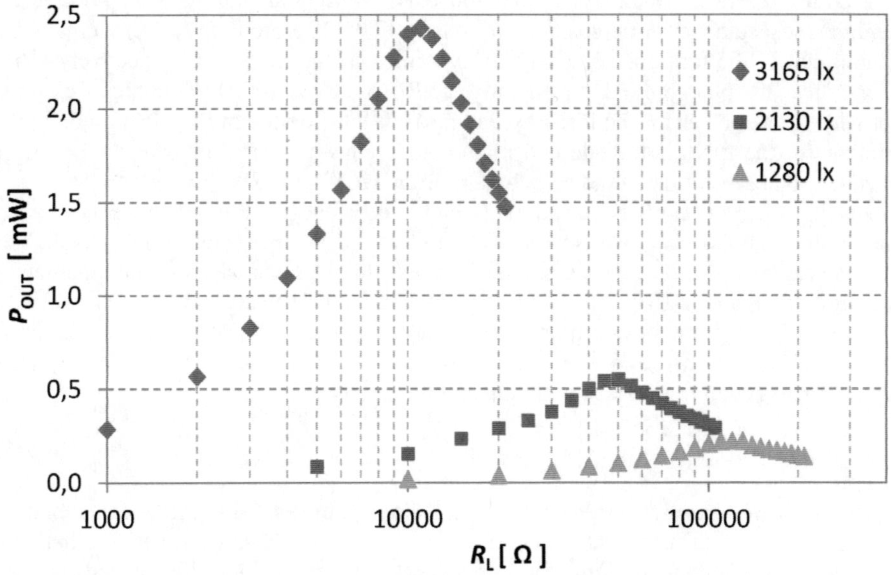

Figure 3. Influence of matching load R_L on the output power P_{OUT} of photovoltaic power converter PPC-6E for different illumination excitation.

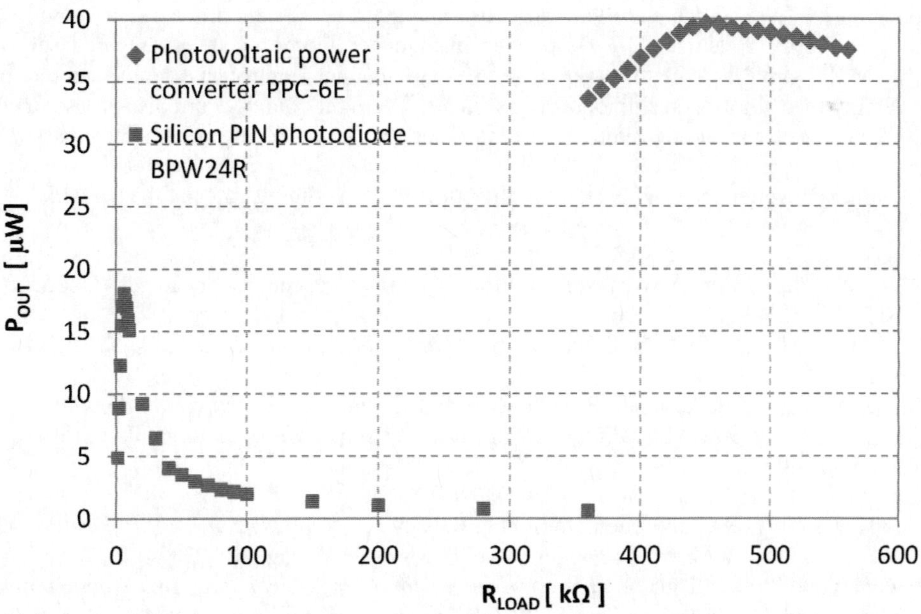

Figure 4. Influence of matching load R_L on the output power P_{OUT} of photovoltaic power converter PPC-6E and Si photodiode for 850nm fiber source.

DC power supplies based on silicon PIN photodiode BPW24R and photovoltaic power converter PPC – 6E are the most suitable for applications in passive FDC. Therefore, the experimental curves of the matching circuit of the transducers have been presented in figures 2 – 4, respectively. The peak output power of the PIN photodiode is in the range of R_{LOAD} below 100 Ω while photovoltaic power converter provides peak power for the loads exceeding 10 kΩ. Influence of illumination on the output power is similar for the PIN photodiode and photovoltaic converter. The ratio of the output power attenuation equals almost 6 for the range of excitation from 3165 lx to 1280 lx.
The measurements (figure 4) confirm considerably higher efficiency of PPC-6E transducer in case of direct excitation through 0.6 mW laser source 850nm from FIS Deluxe Mini OTDR. Available peak power does not exceed 40μW, however, amount of harvested power enough to supply microcontroller system. The tested Photodiode BPW24R cannot be used in the harvesting mode when 0.6 mW laser source is applied. The highest power on the level of 18 μW is still insufficient to cover power requirements of DC/DC converter and the microcontroller. Lower efficiency of the BPW24R transducer is caused by improper coupling interface with optical fiber source.

5. Conclusions

We have built a power supply system for 32-bit microcontroller with ARM-cores and digital thermometer. Four types of converters acting on the basis of electromagnetic radiation were investigated. Each produces enough energy for proper operation of microcontroller. Moreover the output power was so high, that it is possible to expand monitoring system with new sensors. Average time t_{on} needed to excite the microcontroller for almost all tested converters was 3s. Only PPC – 6E illuminated by 0.6 mW laser source 850nm FIS mini OTDR needed 300 s to produce sufficient amount of energy to excite the microcontroller. The best converters proved to be a Si photodiode and a directional antenna RF-PCB. Since our goal was to design the power supply system for FDC, the most convenient converters in our opinion are Si photodiode and PPC – 6E. By using these converters it is possible to build fully autonomous monitoring system in FDC. Despite the much longer t_{on} time, even the standard laser source dedicated to MMF with PPC – 6E still were able to excite the microcontroller, which is another proof of the possibility of realization such monitoring systems. The light coming from distant laser through the optical fiber, can be converted to electric power and supply the monitoring system. Information from monitoring system can be converted in for example serial – to – fiber converter and send back through optical fiber to administrator of the network.

6. References

[1] JDSU website: http://www.jdsu.com/ProductLiterature/ppc6e_ds_pp_ae.pdf [accessed on 10.2015]
[2] Gilbert J and M Balouchi F 2008 *International Journal of Automation and Computing* **5(4)** 334-347
[3] Mitcheson P, Stark B, Yeatman P, Holmes A and Green T 2003 *Analysis and optimisation of MEMS on-chip power supply for self powering of slow moving sensors*, Proc. Eurosensors XVII, pp 48-51
[4] Starner T E 2003 *IEEE Pervasive Computing* **2(4)** 86–88
[5] Roundy S, Steingart D, Frechette L, Wright P, Rabaey J 2004 *Power Sources for Wireless Sensor Networks. Lecture Notes in Computer Science* Springer-Verlag vol **2920** pp 1–17
[6] Mateu L and Moll F 2005 *Review of Energy Harvesting Techniques and Applications for Microelectronics In Proceedings of SPIE – The International Society for Optical Engineering* VLSI Circuits and Systems II SPIE Press vol **5837** pp 359–373
[7] Mahima Arrawatia, Maryam Shojaei Baghini and Girish Kumar 2015 *IEEE Transactions on Antennas and Propagation* **63(4)** 1581-1588
[8] Henrique Saraiva M et al 2014 *Experimental Characterization of Wearable Antennas and Circuits for RF Energy Harvesting in WBANs* 79th IEEE Vehicular Technology Conference

pp 1-5
[9] Mahima Arrawatia, Maryam Shojaei Baghini and Girish Kumar 2015 *Antennas and Wireless Propagation Letters* **PP(99)** 1-4
[10] You K, et al 2011 *900 MHz CMOS RF-to-DC converter using a cross-coupled charge pump for energy harvesting* International Symposium on Radio-Frequency Integration Technology (RFIT) IEEE pp 149-152
[11] Mikio Taguchi et al 2013 *IEEE Journal of Photovoltaics* **4(1)** 96-99
[12] Tatsuya Takamoto, Eiji Ikeda, Hiroshi Kurita and Masamichi Ohmori 1997 *Applied Physics Letters* **70** 381-383
[13] Wolfgang Guter et al 2009 *Applied Physics Letters* **94** 223504-1-3
[14] Mattos Laila S et al 2012 *New Module Efficiency Record: 23.5% under 1-Sun Illumination Using Thin-film Single-junction GaAs Solar Cells* 38th Photovoltaic Specialists Conference (PVSC) ISSN 0160-8371 pp 003187-003190
[15] Jan Gustav Werthen 2008 *Powering Next Generation Networks by Laser Light over Fiber* Optical Fiber communication/National Fiber Optic Engineers Conference OFC/NFOEC print ISBN 978-1-55752-856-8 pp 1-3
[16] Böttger G et al 2008 IEEE Photonics Technology Letters **20(1)** 39-41

Experimental study of energy harvesting in UHF band

Ł Bernacki[1], **R Gozdur and N Salamon**
Department of Semiconductor and Optoelectronic Devices, Lodz University of Technology, Wolczanska 211/215, 90-924 Lodz, Poland

[1]E-mail: lukasz.bernacki@dokt.p.lodz.pl

Abstract. A huge progress of down-sizing technology together with trend of decreasing power consumption and, on the other hand, increasing efficiency of electronics give the opportunity to design and to implement the energy harvesters as main power sources. This paper refers to the energy that can be harvested from electromagnetic field in the unlicensed frequency bands. The paper contains description of the most popular techniques and transducers that can be applied in energy harvesting domain. The overview of current research and commercial solutions was performed for bands in ultra-high frequency range, which are unlicensed and where transmission is not limited by administrative arrangements. During the experiments with Powercast's receiver, the same bands as sources of electromagnetic field were taken into account. This power source is used for conducting radio-communication process and excess energy could be used for powering the extra electronic circuits. The paper presents elaborated prototype of energy harvesting system and the measurements of power harvested in ultra-high frequency range. The evaluation of RF energy harvesters for powering ultra-low power (ULP) electronic devices was performed based on survey and results of the experiments.

1. Introduction

In recent years, the energy harvesting domain has been generating great interests amongst researchers, scientists and engineers. The energy harvesting is defined as the process of extracting small amounts of energy from various sources and its further conversion into electricity [1]. The energy available for harvesting applications is mainly provided by ambient light [2], ambient radio frequency [3], thermal [4] and mechanical sources [5]. Down-sizing efficiency demands and low power consumption trends opened novel research lines on battery recharge via available power sources. The devices that are able to convert dissipated energy into electricity can be used either as battery rechargers in various environments or as the independent power sources [6]. This is particularly relevant to the installation of wireless sensor networks in areas that are inhospitable or difficult to reach. This includes safety monitoring devices, structure embedded micro sensors and medical implants. There are also environmental benefits associated with limiting the disposal of batteries.

2. Scope of energy harvester source

The scope was prepared in the context of available and usable energy sources for harvesting techniques that are still being developed to improve efficiency of energy conversion process. The most popular of these can be classified according to scheme shown in figure 1.

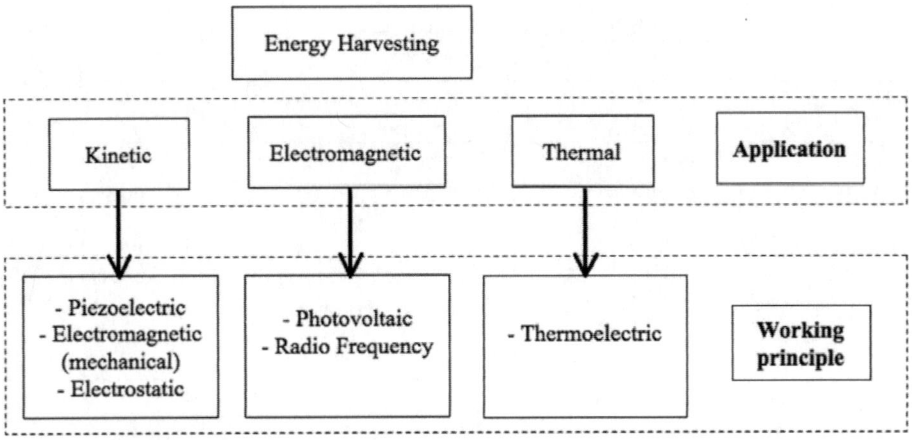

Figure 1. Classification of the main energy harvesting techniques.

For medium and large scale harvesting, the techniques such as solar, wind, hydro and geothermal are well known. Solar energy harvesting is one of the most practical applications due to its higher energy density compared with the other sources. However, it can operate only in the presence of sunlight, which makes the generation of energy during night-time and cloudy day limited. For small scale systems, energy harvesting is used as an alternative energy source to supplement a primary power source [5]. Due to the fact that vibration sources are ubiquitous in the ambient environment, vibration-based energy harvesting is one of the attractive solutions for powering autonomous small-scale microsystems. There is a wide variety of mechanical energy sources including vibrations from industrial machinery and transport, fluid flow, such as air movements, direct human action from walking or in-body motion, such as chest and heart movement, which can be used e.g. for powering pacemakers and orthopedic implants. A great number of these sources is also known and used for large scale power generation (e.g. wind power). However, energy harvesting techniques are mainly focused on a very small scale power generation. Typically, they are used for supplying small electronic devices, where mains or battery power does not provide a viable or convenient solution. Basically, the vibration-to-electricity conversion mechanism can be implemented by piezoelectric, electromagnetic, electrostatic, and magnetostrictive transducers. The piezoelectric transducers are being widely investigated due to their high efficiency and they are commonly used for vibrations. Kinetic energy harvesting is mainly based on the acceleration of a mass. When the base moves, a vibration is set up in the mass-spring system, from which electrical energy can be harvested. On the other hand, the piezoelectric energy harvesters are based on Micro Electro-Mechanical Systems (MEMS) with a few of movable components. Additionally, energy harvester is creating electric power only when the source or base is in motion. Exemplary systems for energy harvesting are listed in table 1.

In contest of energy harvesting, the best way, is to use permanently working sources of energy. Each year, more than 120.000 new base stations around the world, only for the cellular network, are deployed [6]. This significant increase of the wireless networks results in a rise of Radio Frequency (RF) power in the form of electromagnetic waves in the environment. Nonetheless, only a small amount of RF power is used for communication and the rest is wasted by dissipation of heat or absorption in other materials. Wasted RF energy can be used as a green source of energy, and RF energy harvesting system powered from ambient can be applied as a power supply of various ultra-low power electronics devices. The electrical power generated by RF energy harvesting system is usually low therefore, depending on technique, it is possible to supply only ULP devices in steady way. For example one can supply wireless sensor network in medical application, battery systems and radio frequency identification (RFID) systems [12].

Table 1. Typical energy harvesting sources.

Energy source	Condition	Performance	Parameters	Elaboration
Solar GaAs	One Sun 1000 W/m^2	33.3 mW	0.9927 cm^2	[Kayes et al. 2011]
Solar InGaAsSb	Infrared	200 mW	-	[Choi et al. 2013]
	4.76 Hz	58 μW	780 g	
Electrostatic Al/poliester	6.0 Hz	36 μW	1200 g	[Tashiro et al. 2002]
Thermal	ΔT = 5 °C	60 μW/cm^2	1.12 mm^2	[Mathna et al. 2009]
Thermal	ΔT = 10 °C	135 μW/cm^2	-	[Mathna et al. 2009]
				[Shenck 1999]
Piezoelectric PZT	0.91 Hz	8400 μW	37161 mm^3	[Shenck et al. 2001]

Table 2. Typical RF energy harvesting sources.

Antenna type	Bandwidth [MHz]	Power [μW]	Gain [dBi]	Elaboration
Dipole	81.9-84.7	10	1.37-1.83	[Noguchi 2013]
Dipole	DTV-3G	3.1	4.5	[Piñuela et al. 2013]
Dipole	868	1.9	0.659	[Sun 2013]
PCB	955	2.5	-	[Stoopman 2014]
Yagi-Uda	900, 2100	3.1	9.9-13.3	[Din 2012]
Patch	3G	398	-	[Ahn 2014]
Patch fractal x9	2400	3.1	4.5	[Olgun 2012]

RF energy harvesting is receiving increasing interest as evidenced by the growing number of scientific papiers. There are not many reports showing accurate measurements and descriptions of the systems, complete data and examples of practical applications. For this reason, the authors have attempted to develop a position and to perform the measurements that show real results.

3. RF energy harvester setup

The application of harvested RF power coming from the ambient environment should cope with two real-life challenges: low RF power density and limited computational capabilities. The problem related to small amount of RF radiated power can be alleviated by storing the harvested energy in battery cell or even in supercapacitor. Obviously, this is a slow and time-sensitive process. The second issue implies that merely simple operations can be reasonably expected from the harvested power, which may perfectly concur with the power consumption profile of wireless sensors. This means that for current applications, only powering low duty cycle devices is meaningful. Moreover, the size of an RF energy harvester is usually designed to be small as compared to other harvestable sources of energy (thermal, mechanical, etc.). It is well known that an antenna or antenna array can be enlarged and designed to coherently capture more wireless energy, which could present an approach to enhancing the received power. However, it is not appropriate when the size of the power harvester is too large to be practical. Elaborated prototype of energy harvesting system consists of transmitter, designed receiving antenna, the rectification unit, step-up converter and the load. The block diagram of the setup is shown in figure 2.

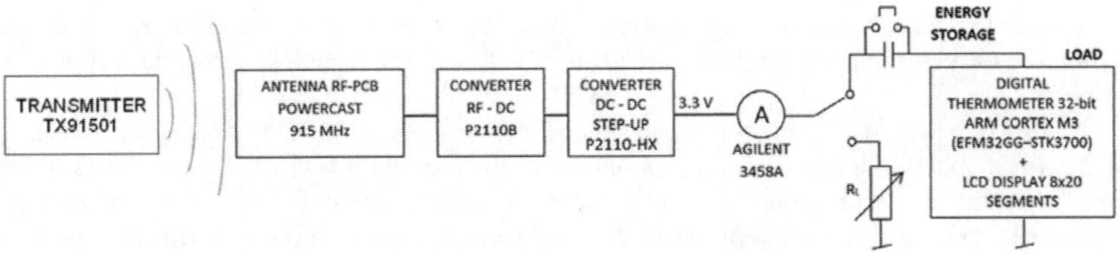

Figure 2. Block diagram of measurement energy harvesting system.

4. Results and discussion

A first step in designing the RF energy harvesting system is to select the frequencies at which power will be harvested.

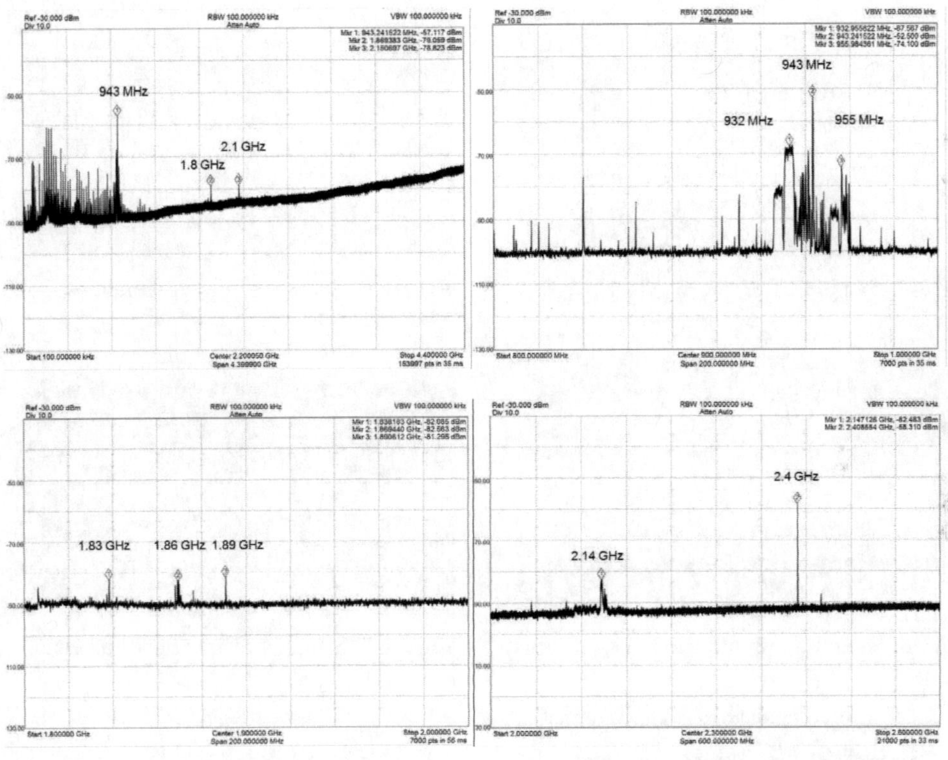

Figure 3. The measured power densities in a typical ambient environment. The measurement method is not accurate enough to provide a quantitative information, but these data are representative of what spectra can be expected in an urban environment.

The wireless spectrum shown in figure 3, is full of signals with different frequencies and power levels, ranging from cellular standards, WLANs and TV signals. The criteria that control the selection of certain frequencies for the purpose of energy harvesting are wide deployment and power level. Obviously, due to multipath electromagnetic wave propagation effects, nonuniform base-station grids and various human-related movements, RF power density is not always constant and uniform over space and time. GSM 900/1800 and DTV signals cover most of the world expect for Europe [20] (as shown in figure 2). Thus it is almost guaranteed that GSM signals would be available regardless of where the harvesting system is placed. Additionally, narrow band signals are appropriate in connection with high quality factor (high-Q) harvester circuits, and they are available all the time, which is highly desirable for most of the applications.

In the next step we designed five simple antennas based on standard types. The antennas were designed with a radiator printed on the bottom of a dual-layer FR4 substrate with thicknesses of 0.5 mm and 1 mm, in which ground plane lies at the bottom of the structure. On top of FR4 dipole (1 dBi, figure 9, 11) and patch (6.1 dBi, figure 10) structure of the antennas were printed. Additionally, microstrip (figure 8) [17, 20, 21], triple bands (0,3 dBi, figure 7) [22] and three dipole windings 1/4/8 turns (figure 4, 5, 6, respectively) antenna with 0.2 mm wire [23] were designed. This structures are very simple but they are used and described in the scientific papers.

Figure 4. Sample of dipole winding / 1 turn antenna.

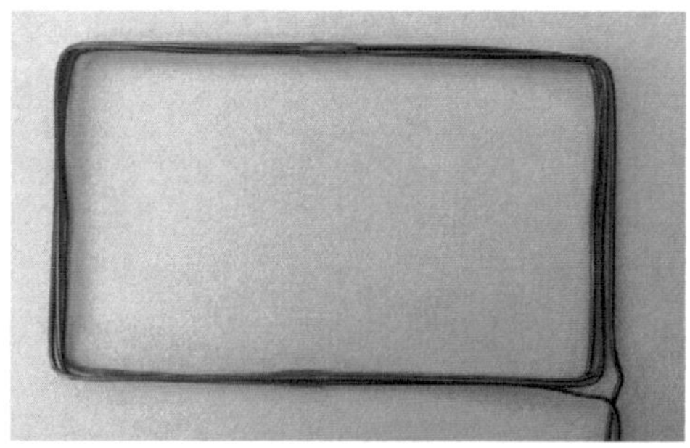

Figure 5. Sample of dipole windings / 4 turns antenna.

Figure 6. Sample of dipole windings / 8 turns antenna.

Figure 7. Sample of triple bands antenna.

Figure 8. Sample of microstrip antenna.

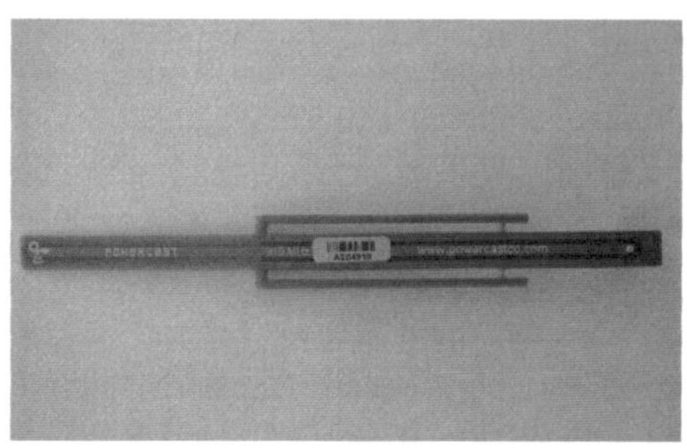

Figure 9. Sample of dipole antenna.

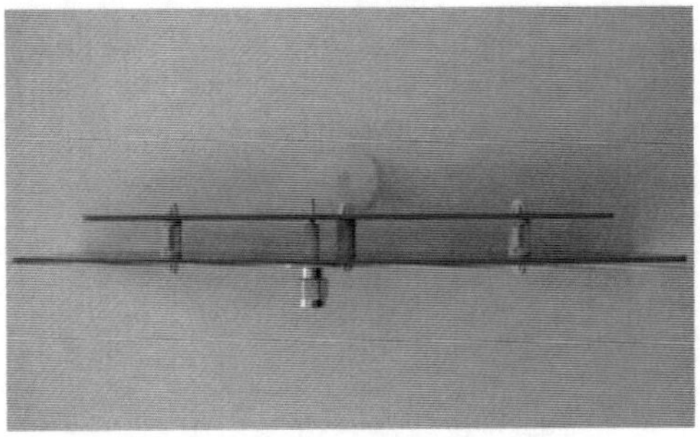

Figure 10. Sample of patch antenna.

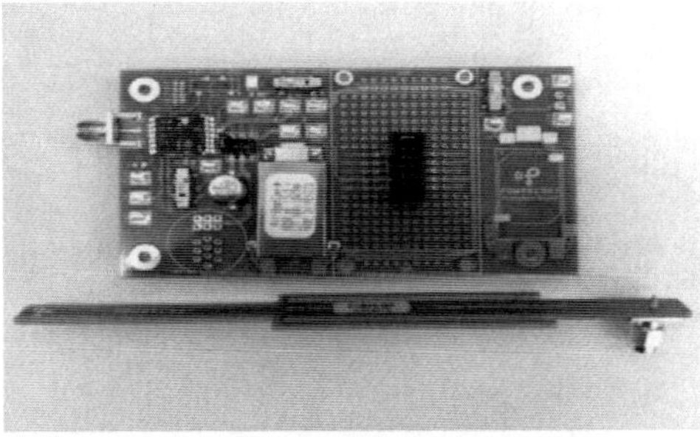

Figure 11. Radio Frequency Energy Harvesting system with dipole antenna.

The impedance matching network is the most critical element in RF energy harvesting system and it has to be designed very carefully for optimal output. The matching circuit, to match the load impedance to impedance of the source (antenna in this case), is designed for 50 Ω and is tuned to 915 MHz. Harvested energy can be stored or transmitted to the step-up DC-DC converter P2110-HX with adjusted output voltage at level 3.3 V. Powercast TX91501 transmitter that provides a maximum power of 3 W EIRP (Equivalent or Effective Isotropic Radiated Power) in the UHF band around 915 MHz, was used as temporary and reference power source. This frequency was chosen because it is representing data uplink from mobile phone to GSM station in communication between them and does not violate European restriction (according to the Nordic Mobile Telecommunication System). The measurements were performed with an application of high accuracy multimeter Agilent 3458A. In order to determine the operating conditions, efficiency and tracking of maximum power output of the harvester, the demo board DC2080A with 32bit ARM CORTEX M3 microprocessor was applied as the load. The received power P_R at a given distance d was calculated from Friis equation:

$$P_R = P_0 G_0 G_R \left(\frac{\lambda}{4\pi d}\right)^2 \qquad (1)$$

Where G_0 and G_R are the antenna gains (with a respect to an isotropic radiator) of the transmitting and receiving antennas, respectively and λ is the wavelength.

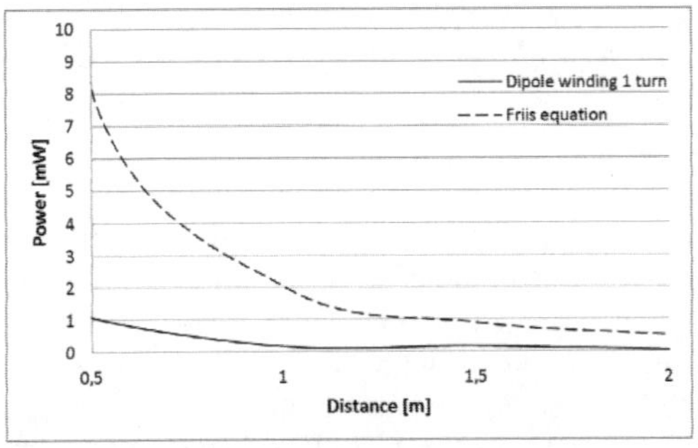

Figure 12. Sample of dipole winding / 1 turn antenna.

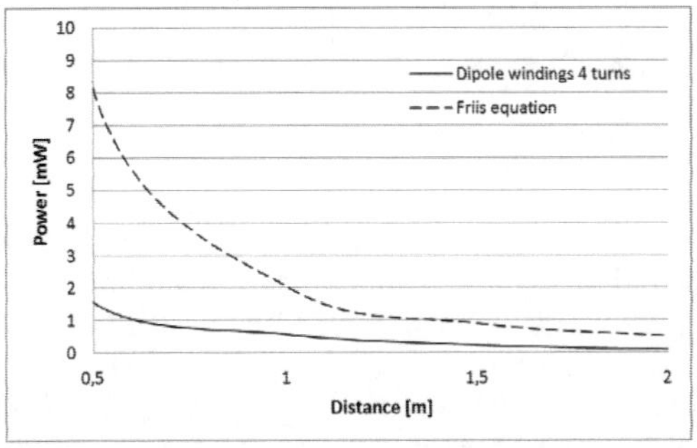

Figure 13. Sample of dipole winding / 4 turn antenna.

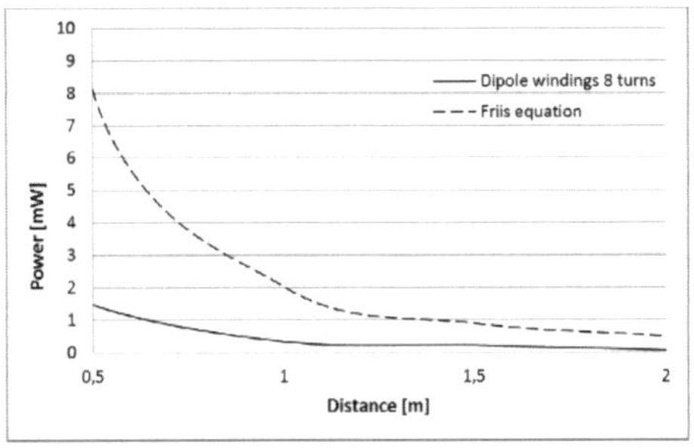

Figure 14. Sample of dipole winding / 8 turn antenna.

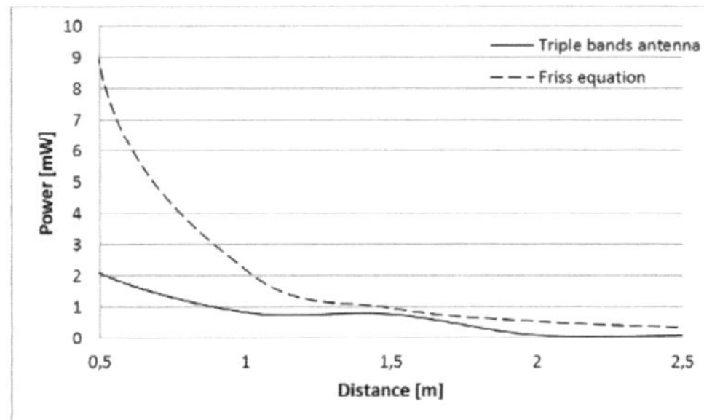

Figure 15. Sample of triple bands antenna.

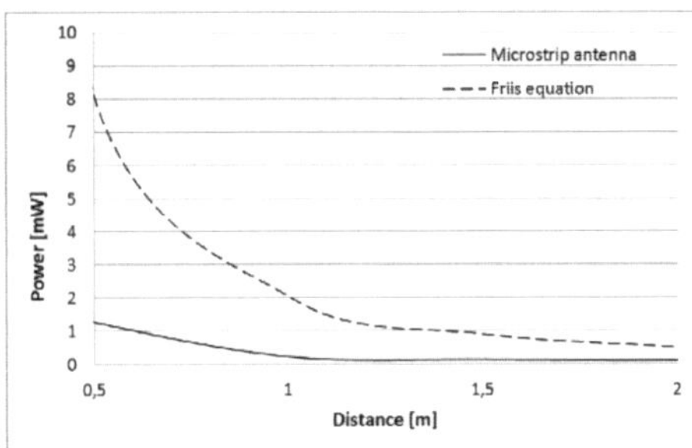

Figure 16. Sample of microstrip antenna.

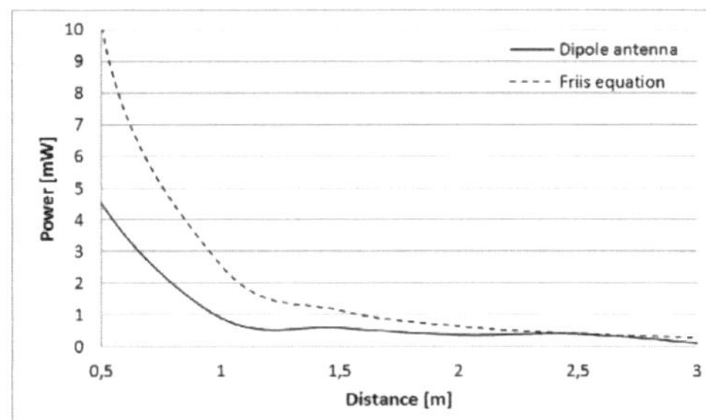

Figure 17. Sample of dipole antenna.

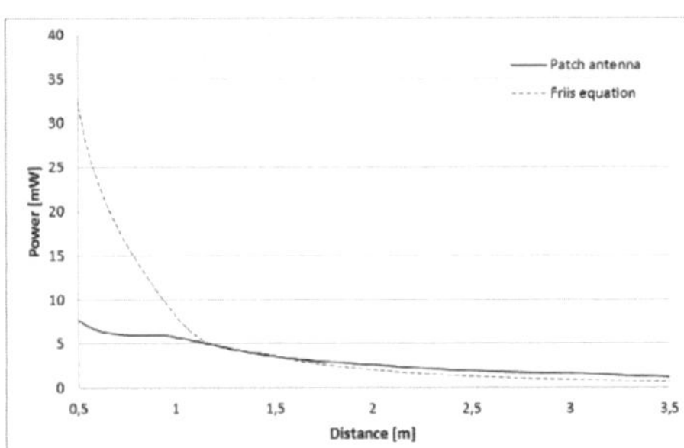

Figure 18. Sample of patch antenna.

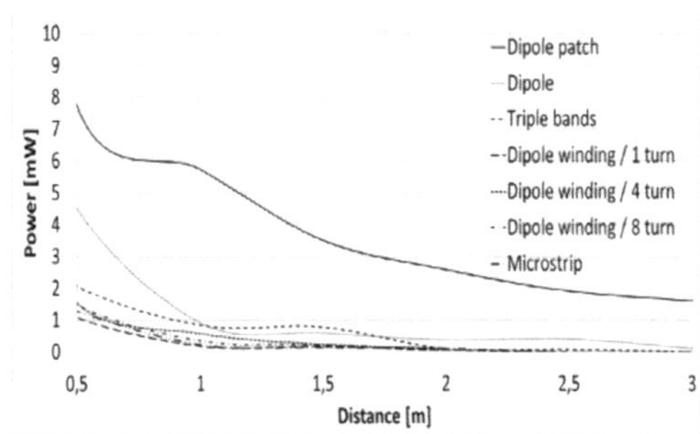

Figure 19. The measured power results obtained for different types of antennas versus distance.

Graph presented in figure 19 shows power gained from the energy harvesting system working in band with 915 MHz center frequency. The highest values of power were obtained at the distance of 0.5 m and they reached 7.77 mW for dipole patch antenna with 6.1 dBi gain and 1400 Ω load resistance R_L. It was enough to wake up the processor from deep sleep mode to active mode in time t_{ON} around 0.8 s. All the tests proved that described energy harvester can work as a power supply and maintain continuously an active state of demo board using the energy harvested from ambient. Table 3 summarizes the performance of the voltage multiplier and RF energy harvesting system.

Table 3. Test results of elaborated antennas at distance 0.5 m in band with 915 MHz center frequency.

Antenna type	R_L [kΩ]	I [mA]	P [mW]	t_{ON} [s]
Dipole patch	1.4	1.74	7.77	0.8
Dipole	2.4	1.38	4.54	1.5
Triple bands	5.2	0.63	2.10	25.5
Dipole winding / 1 turn	10.0	0.33	1.09	3.0
Dipole winding / 4 turn	7.0	0.47	1.55	10.2
Dipole winding / 8 turn	7.3	0.45	1.49	17.1
Microstrip	8.5	0.39	1.28	3.4

Further tests, during which the mobile phone working in EGSM band played the role of a power source, were performed for antenna with the highest value of gained power. Table 4 summarizes the performance of the voltage multiplier and RF energy harvesting system.

Table 4. Test results of elaborated antennas at distance 0,5m in EGSM band.

Antenna type	R_L [kΩ]	I [mA]	P [mW]	t_{ON} [s]
Dipole patch	40	83	0.27	230
Dipole	110	30	0.10	650
Triple bands	250	20	0.04	1600

5. Conclusions

Presented results prove that the electromagnetic energy, which is naturally abundant in many practical applications, can be scavenged and channelled to a rechargeable energy storage device (e.g., a battery). By applying low-power and duty-cycling techniques to the loading system the harvested energy can viably replenish the total energy consumed by the system, potentially extending its operational life indefinitely without manual/external recharge or battery replacement cycles, which may be otherwise prohibitive in applications such as remote wireless microsensors and bio implantable devices.

6. References

[1] Pillai M A and Deenadayalan E 2013 *Int. J. of precis. eng. man.* **15** 949–965

[2] Kayes B M, Nie H, Twist R, Spruytte S G, Reinhardt F, Kizilyalli I C and Higashi G S 2011 *Proc. of the 37th IEEE Photovoltaic Spec. Conf.*

[3] Hong H, Tong S J, Cai X, Shi X and Zhu X 2012 *Microwave and Millimeter Wave Technology (ICMMT) IEEE* **5** 1–4

[4] Mathna C, Donnell T O, Martinez – Catala R V, Rohan J and Flynn B O 2008 *Elsevier: Talanta* **75** 613–623

[5] Caliò R, Rongala U B, Camboni D, Milazzo M, Stefanini C, Petris G and Oddo C M 2014 *Sensors* **14** 4755–4790

[6] Oha S, Ramasamya M and Varadan V K 2014 *Proc. of SPIE* **9060** 90601B–1

[7] Choi S and Sarabandi K, 2013 *Antennas and Propagation Society Int. Symp. (APSURSI) IEEE* 146–147

[8] Tashiro R, Kabei N, Katayama K, Tsuboi F and Tsuchiya K 2002 *J. Artif. Organs* **5(4)** 239–45

[9] Ramesh G P and Rajan A 2014 *Int. Conf. on Communication and Signal Processing (India)*

[10] Shenck N S 2009 *M. S. thesis* MIT Cambridge MA.

[11] Shenck N S and Paradiso J A 2001 *IEEE Micro* **21** 30–42

[12] Jankowski – Mihułowicz P, Kalita W, Skoczylas M and Węglarski M 2013 *International Journal of Antennas and Propagation* **2013** 242840

[13] Noguchi A. and Arai H., *Asia-Pacific Microwave Conference Proceedings* **2013** 86-88
[14] Piñuela M, Mitcheson P D and Lucyszyn S 2013 *IEEE Transactions on microwave theory and techniques* **2013 61** 7
[15] Sun H., Guo Y., He M. and Zhong Z. *Antennas and Wireless Propagation Letters* **2013 12** 918-921
[16] Stoopman M., Keyrouz S., Visser H. J., Philips K. and Serdijn W. A. IEEE Journal Of Solid-State Circuits **2014 49** 3
[17] Din N M, Chakrabarty C K, Bin Ismail A, Devi K K A and Chen W Y 2012 *Progress In Electronic Research* **132** 49–69
[18] Ahn C.J., IEEE *ISCE*, **2014** 1-2
[19] Olgun U., Chen C. C. and Volakis J. L., *IET Microw. Antennas Propag,* **2012 6** 1200-1206
[20] Aminov P and Agrawal J P 2014 *Electronic Components and Technology Conf.* 1838–1841
[21] Tudose D Ş and Voinescu A 2013 *19th Int. Conf. on Control Systems and Computer Science* 184–188
[22] Pham B L and Pham A V 2013 *IEEE MTT-S Microwave Symp. Digest (IMS)* 1–3
[23] Beng Lim T, Meng Lee N and Kiat Poh B 2013 *IEEE IMWS-BIO* 1–3

The influence of core material on transient thermal impedances in transformers

K Górecki and K Górski

Gdynia Maritime University, Department of Marine Electronics, Morska 83, Gdynia, Poland

E-mail: k.gorecki@we.am.gdynia.pl

Abstract. In the paper the results of measurements of thermal parameters of impulse-transformers containing cores made of different ferromagnetic materials are presented. Investigations were performed with the use of methods worked out in Gdynia Maritime University. The obtained results of measurements prove that the material of the core does not influence transient thermal impedance of the winding, whereas this parameter visibly changes with the change of spatial orientation of the transformer. In turn, the material of the core decides about transient thermal impedance of the core. Additionally, the influence of the core material on temperature distribution on the surface of the transformer was analysed.

1. Introduction

Impulse-transformers are commonly used in switch-mode power supplies [1, 2, 3]. The considered elements have a simple construction - they consist of the ferromagnetic core and windings. The properties of both these components depend on temperature, whose change causes changes in the value of exploitive parameters of the core and windings [4, 5, 6]. Particularly, if the core temperature is higher than the Curie temperature, permeability of this core decreases to 1, and when the windings temperature is higher than its admissible value, isolation of wires can be destructed [1, 2].

The temperature of the core and windings of the transformer during its operation is higher than the ambient temperature due to self-heating phenomena in the core and in the windings, as well as the mutual thermal coupling between these components of the transformer [4, 7, 8, 9]. In the papers [4, 8, 9] compact thermal models of the transformer are proposed. These models use the idea of the transformer's own and mutual transient thermal impedances well-known from models of semiconductor devices [10, 11, 12]. As it is known from some papers, e.g. [11, 13], thermal parameters of semiconductor devices depend on such factors as dissipated power, dimensions of the considered devices and construction of the cooling system. Therefore, it can be expected that thermal parameters of transformers depend on their dimensions and parameters of materials used to produce ferromagnetic cores. Magnetic materials are characterized by different values of thermal conductance, which should influence transient thermal impedance of the transformer.

In the paper the results of measurements of transformers' own and mutual transient thermal impedances, obtained with the use of the measurement method elaborated at Gdynia Maritime University [14], are presented. These transformers contain cores made of different materials. Additionally, inequalities of temperature distribution on the surface of the investigated elements are discussed.

2. Measurement methods

In the research the method of measurements of the transformer's own and mutual transient thermal impedances described in the paper [14] is used. This method is realised in the measurement set presented in Figure 1.

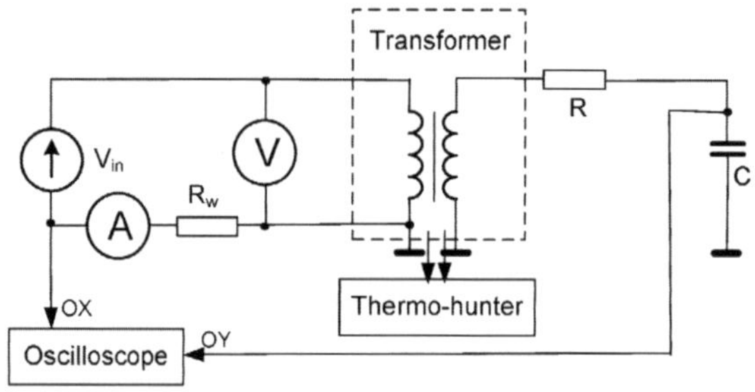

Figure 1. The measurement set to measure thermal parameters of transformers [14]

The measurements are conducted in two steps. The first step needs stimulations of the primary winding with a jump of the current and the measurement of temperature changes of windings and of the core by means of the thermo-hunter until the thermally steady-state is obtained. These measurements are used to calculate transient thermal impedance of the winding $Z_{thU}(t)$ and mutual transient thermal impedance between the core and the windings $Z_{thUR}(t)$ using the following formulas:

$$Z_{thU}(t) = \frac{T_U(t) - T_a}{P} \quad (1)$$

$$Z_{thUR}(t) = \frac{T_R(t) - T_a}{P} \quad (2)$$

where $T_U(t)$ and $T_R(t)$ denote waveforms of the winding and core temperatures, respectively, T_a is the ambient temperature, whereas P denotes power dissipated in the winding, which is equal to the product of the winding current and the voltage on the primary winding.

In the second step, the primary winding of the transformer is stimulated by a sinusoidal signal of frequency f and the temperature of the core is measured by the thermo-hunter. When the steady state is obtained, in the moment $t = 0$ the power supply of the primary winding is switched off and waveforms of temperature of the core and windings are measured. On the basis of the area S_H of the obtained hysteresis loop $B(H)$ of the core and the measured waveform of the core temperature, transient thermal impedance of the core $Z_{thR}(t)$ is calculated using the following formula

$$Z_{thR}(t) = \frac{T_R(t=0) - T_R(t)}{V_R \cdot f \cdot S_H} \quad (3)$$

where V_R represents the volume of the core.

The detailed description of the method is included in [14].

3. Measurement results

Using the method presented in section 2, the measurements of thermal parameters of transformers containing toroidal cores of the diameter equal to about 26 mm are performed. The core made of powdered iron (RTP), the ferrite core (RTF) and the nanocrystaline core (RTN) are applied. Each of the considered transformers has two windings made of 30 turns of copper wire in enamel of the diameter equal to 0.8 mm.

In the further part of this section the results of measurements illustrating the influence of core material and spatial orientation of the transformer on the courses of transient thermal impedances $Z_{thU}(t)$, $Z_{thR}(t)$ and $Z_{thUR}(t)$ are presented. The spectrum of transient thermal impedances for all the considered cooling conditions and selected temperature distributions on the surface of the investigated transformers are also shown. All the measurements are performed at the constant ambient temperature equal to 22°C. In all the figures presented in this section, solid lines correspond to the transformer situated horizontally, and dashed lines - the transformer situated vertically. With the red colour the measured courses of $Z_{thU}(t)$ are marked, with the blue colour - courses $Z_{thR}(t)$, and with the black colour - courses $Z_{thUR}(t)$.

The spectrum of transient thermal impedances illustrates the values of parameters describing waveforms of $Z_{th}(t)$ by means of the classical analytic formula [10, 12, 15]

$$Z_{th}(t) = R_{th} \cdot \left[1 - \sum_{i=1}^{N} a_i \cdot \exp\left(-\frac{t}{\tau_{thi}}\right)\right] \tag{4}$$

where R_{th} denotes thermal resistance, N – number of thermal time constants τ_{thi} corresponding to coefficients a_i.

In Figure 2a the measured waveforms of transient thermal impedances $Z_{thU}(t)$, $Z_{thUR}(t)$ and $Z_{thR}(t)$ for the transformer containing the nanocrystaline core are presented, whereas in Figure 2b – the spectrum of thermal time constants of this transformer is shown. Parameters values of the thermal model of the considered transformer are collected in Table 1.

a) b)

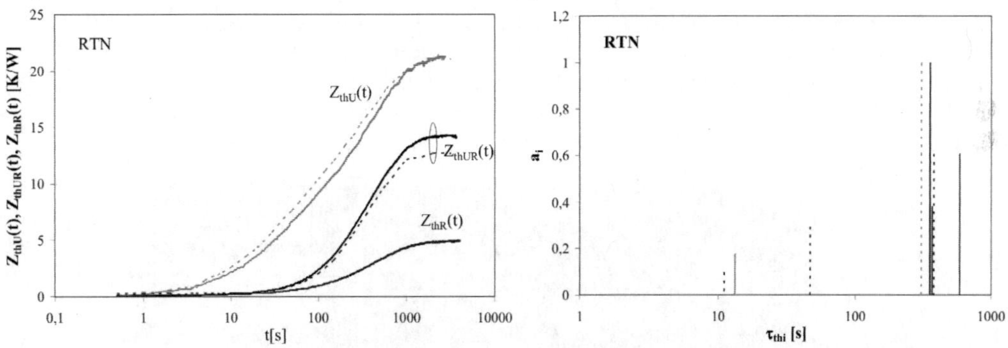

Figure 2. Transient thermal impedances in the transformer with the RTN core (a) and the spectrum of thermal time constants of this transformer(b)

Table 1. Parameters values of the thermal model of the transformer with the RTN core.

parameter	transformer situated horizontally			transformer situated vertically	
	$Z_{thU}(t)$	$Z_{thUR}(t)$	$Z_{thR}(t)$	$Z_{thU}(t)$	$Z_{thUR}(t)$
R_{th} [K/W]	20.94	14.1	4.91	21.02	12.56
a_1	0.826	1	0.606	0.608	1
a_2	0.174		0.382	0.294	
a_3			0.012	0.098	
τ_{th1} [s]	359.1	361.7	597.9	385.4	310.1
τ_{th2} [s]	13.35		373.6	47.3	
τ_{th3} [ms]			40	98	

As one can notice in Figure 2a, the process of heating the core and winding of the transformer runs slowly. The indispensable time to obtain the steady state exceeds 3000 s. It is worth paying attention to the fact that the process of heating the winding runs more quickly, and the courses $Z_{thUR}(t)$ and $Z_{thR}(t)$ are late with regard to the course $Z_{thU}(t)$ by even about 100 s. Additionally, it is visible that the steady-state values of transient thermal impedance of the winding are even about 20% higher than the values of transient thermal impedance between the winding and the core of this transformer. On the other hand, the values of transient thermal impedance of the core are even four times smaller than $Z_{thU}(t)$. The influence of orientation of the transformer in the vertical-line or in the horizontal-line on the course of transient thermal impedance is visible only in the case of $Z_{thUR}(t)$, where the value of this parameter at the steady-state at vertical orientation is about 10% lower than at horizontal orientation of this element.

In Figure 2b it is visible that the presented in Figure 2a waveforms of transient thermal impedances can be described with the use of 1 to 3 thermal time constants, whereas the prevailing thermal time constant accepts values in the range from 200 to 300 s. Orientation of the transformer does not influence in an essential manner the value of thermal time constants.

In Figure 3 distribution of temperature on the surface of the investigated transformer with the RTN core, obtained at the steady-state at different conditions of power supply of this transformer, are shown. At dc stimulation the current of the primary winding is equal to about 9 A, whereas at the stimulation of the primary winding with the sinusoidal current the amplitude is 2.4 A and frequency 5.5 kHz.

As one can notice for the transformer with the RTN core, at the stimulation with the direct current, temperature on its surface at horizontal orientation accepts the values in the range from 40°C to 78°C, at vertical orientation - the values of temperature from 40°C to 71°C, and at the stimulation of the transformer with the sinusoidal current, temperature on its surface at horizontal orientation has the values in the range from 30°C to 42°C. It should be noted that visible differences between the temperature of the core and winding occur. At the power supply with the direct current the winding has higher temperature, and at the power supply with the sinusoidal current - the core. Warmer areas of the windings show the visible difference of temperatures not higher than several Celsius degrees, similarly to the values of temperature on the surface of the core.

a) b) c)

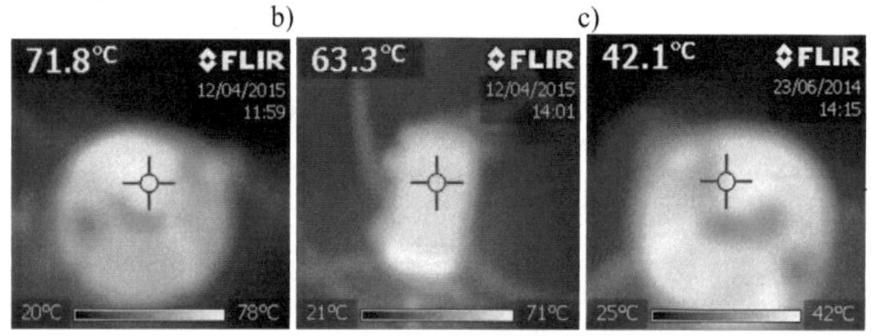

Figure 3. Temperature distribution on the surface of the transformer with the RTN core at the stimulation by: a) the dc current at horizontally situated transformer, b) the dc current at the vertically situated transformer, c) the sinusoidal waveform of the current at the horizontally situated transformer

In Figure 4a the measured waveforms of transient thermal impedances $Z_{thU}(t)$, $Z_{thUR}(t)$ and $Z_{thR}(t)$ for the transformer containing the powder core (RTP) are presented, whereas in Figure 4b – the spectrum of thermal time constants of this transformer is shown. Parameters values of the thermal model of the considered transformer are collected in Table 2.

a) b)

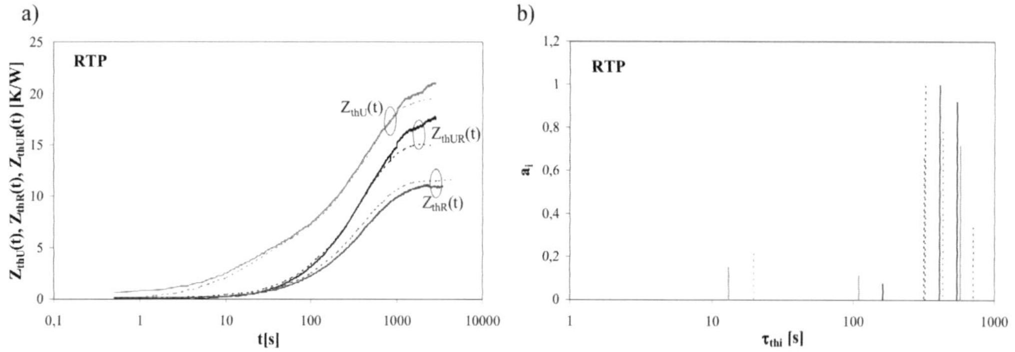

Figure 4. Transient thermal impedances in the transformer with the RTP core (a) and the spectrum of thermal time constants of this transformer(b)

Table 2. Parameters values of the thermal model of the transformer with the RTP core.

parameter	transformer situated horizontally			transformer situated vertically		
	$Z_{thU}(t)$	$Z_{thUR}(t)$	$Z_{thR}(t)$	$Z_{thU}(t)$	$Z_{thUR}(t)$	$Z_{thR}(t)$
R_{th} [K/W]	20.98	17.62	10.94	19.36	14.82	11.69
a_1	0.717	0.922	1	0.784	1	0.341
a_2	0.113	0.078		0.216		0.659
a_3	0.153					
a_4	0.017					
τ_{th1} [s]	572.8	541.9	410.2	433.2	325.9	708.6
τ_{th2} [s]	109.3	161.9		19.57		320
τ_{th3} [s]	13.06					
τ_{th4} [ms]	40					

As one can notice in Figure 4a the process of heating the core and the winding of the transformer with the RTP core occurs similarly as for the transformer with the RTN core. The time indispensable to obtain the steady state exceeds 3000 s. The obtained value $Z_{thU}(t)$ at the steady-state amounts to about 22 K/W and it is practically the same as for the transformer with the RTN core, whereas values $Z_{thUR}(t)$ and $Z_{thR}(t)$ for the transformer with the RTP core are considerably (even twice) higher than for the transformer with the RTN core. At vertical orientation smaller by about 10 - 20 % values of $Z_{thU}(t)$ and $Z_{thUR}(t)$ than for horizontal orientation of this transformer are obtained. In turn, the influence of orientation of the transformer on the course $Z_{thR}(t)$ is omittably weak.

In Figure 4b it is visible that the presented in Figure 4a waveforms of transient thermal impedances can be described with the use from 1 to 3 thermal time constants, whereas the prevailing thermal time constant accepts values in the range from 200 to 500 s. It is visible that at vertical orientation of the transformer deterioration of the prevailing thermal time constant by even about 50% is observed.

In Figure 5 distribution of temperature on the surface of the investigated transformer with the RTP core obtained at the steady state at different conditions of stimulation of this transformer are shown. At dc stimulation the current of the primary winding is equal to about 9 A, whereas at the stimulation of the primary winding with the sinusoidal current the amplitude is 2.4 A and frequency 5.5 kHz.

a) b) c) d)

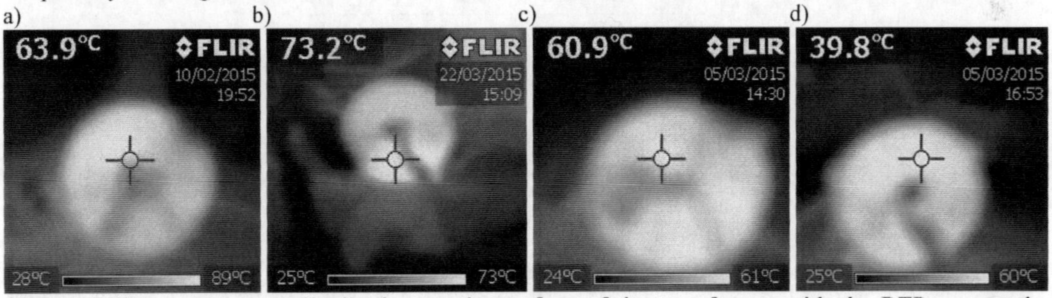

Figure 5. The temperature distribution on the surface of the transformer with the RTP core at the stimulation by: a) the dc current at the horizontally situated transformer, b) the dc current at the vertically situated transformer, c) the sinusoidal waveform of the current at the horizontally situated transformer, d) the sinusoidal waveform of the current at the vertically situated transformer

As one can notice for the transformer with the RTP core, at the stimulation with the direct current, temperature on its surface at horizontal orientation accepts values in the range from 40°C to 89°C, at vertical orientation - values of temperature in the range from 40°C to 73°C. In turn, at the stimulation of the transformer with the sinusoidal current, temperature on its surface at vertical and horizontal orientation accepts values of temperature in the range from 40°C to 61°C.

It is proper to notice that visible differences between the temperature of the core and the winding appear. However, warmer areas of windings show not big differentiation in temperature, not exceeding several Celsius degrees, similarly to values of temperature on the surface of the core.

The measurements of temperature distribution on the surface of the transformer and the courses of transient thermal impedances are performed also for transformers containing the ferrite core (RTF). The results of such measurements are shown in Figure 6 for the transformer situated horizontally. As it is visible, the obtained results qualitatively agree with the presented above results of measurements of transformers with the RTP and RTN cores. Parameters values of the thermal model of the transformer with the RTF core are collected in Table 3.

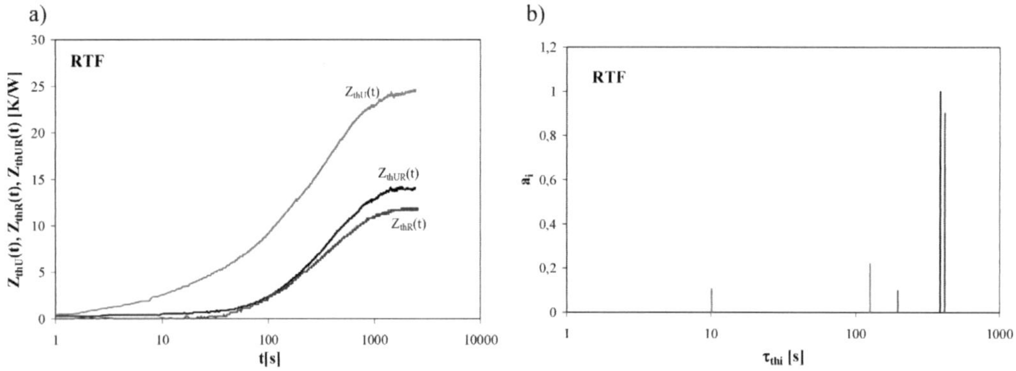

Figure 6. Transient thermal impedances in the transformer with the RTF core (a) and the spectrum of thermal time constants of this transformer(b)

Table 3. Parameters values of the thermal model of the transformer with the RTF core situated horizontally.

parameter	transformer situated horizontally		
	$Z_{thU}(t)$	$Z_{thUR}(t)$	$Z_{thR}(t)$
R_{th} [K/W]	24.55	14	11.9
a_1	0.669	1	0.9
a_2	0.221		0.1
a_3	0.105		
a_4	0.005		
τ_{th1} [s]	415.8	384.35	413.6
τ_{th2} [s]	125.1		195.4
τ_{th3} [s]	10.1		
τ_{th4} [ms]	40		

The obtained values of thermal resistance of the winding is equal to about 25 K/W, the mutual thermal resistance between the winding and the core is equal to about 14 K/W and thermal resistance of the core is equal to about 12 K/W. Thermal time constants accept values in the range from 10 s to about 400 s. Therefore, time indispensable to obtain the steady state is shorter than for the other considered transformers.

The temperature distribution on the surface of considered transformer are also measured and the obtained results are similar to temperature distributions presented in Figure 5 for the RTP core.

4. Conclusions

In the paper the results of measurements of transformers' own and mutual transient thermal impedances in transformers containing cores made of different materials and temperature distribution on the surface of these elements at the steady-state are presented. From the obtained results of measurements it results that the material of the core has a visible influence on the waveforms of transient thermal impedances of the core included in the transformer, but it influences transient thermal impedance of the winding in an omittably weak way. The highest value of this transient thermal impedance is the highest for the transformer with the RTF core. Differences in the waveforms of transient thermal impedance of the core can be a result of thermal conductance of the core material. The highest values of the transient thermal impedance of the core is obtained for transformer with the RTF core. In the steady state its value is even twice higher than the value of this parameter for the transformer with the RTN core. On the other hand, the influence of transformers orientation on their transient thermal impedances for the transformer situated vertically is visible, and typically smaller values of these parameters are obtained.

The obtained distribution of the surface temperature of the transformer shows that inequality of distribution of temperature in the examined transformers does not exceed a dozen or so kelvins. This justifies the use of compact thermal models in the description of thermal properties of the examined transformers.

5. References

[1] Barlik RJ, Nowak KM 2014 *Energoelektronika. Elementy podzespoły, układy* (Warszawa: Oficyna Wydawnicza Politechniki Warszawskiej)

[2] Ericson R, Maksimovic D 2001 *Fundamentals of Power Electronics* (Norwell: Kluwer Academic Publisher)

[3] Rashid MH 2007 *Power Electronic Handbook* (Academic Press, Elsevier)

[4] Górecki K, Rogalska M 2014 *Microelectronics Journal* **45** (12) 1795-1799

[5] Wilson PR, Ross JN, Brown AD 2002 *IEEE Transactions on Power Electronics* **17** (1) 55-65

[6] Van den Bossche A, Valchev VC 2005 *Inductors and Transformers for Power Electronics* (Boca Raton: CRC Press, Taylor & Francis Group)

[7] Górecki K, Detka K, Zarębski J 2013 Pomiary wybranych parametrów i charakterystyk materiałów i elementów magnetycznych *Elektronika* **1** 18-22

[8] Górecki K, Zarębski J 2009 *Microelectronics Reliability* **49** (4) 424-430

[9] Górecki K, Rogalska M, Zarębski J 2014 *Microelectronics Reliability* **54** (5) 978-984

[10] Janke W 1992 *Zjawiska termiczne w elementach i układach półprzewodnikowych* (Warszawa: WNT)

[11] Górecki K, Zarębski J 2014 *IEEE Transactions on Components Packaging and Manufacturing Technology* **4** (3) 421-428

[12] Górecki K, Zarębski J 2010 *IEEE Transactions on Components and Packaging Technologies* **33** (3) 643-647

[13] Oettinger FF and Blackburn DL 1990 Semiconductor measurement technology: thermal resistance measurements *U. S. Department of Commerce* **NIST/SP-400/86**

[14] Górecki K, Zarębski J, Detka K, Rogalska M 2013 Sposób i układ do pomiaru własnych wzajemnych rezystancji termicznych elementu indukcyjnego *European Patent Application* **EP 13460073**

[15] Szekely V 1997 *Microelectronic Journal* **28** (3) 277-292

Analysis of normal zone propagation in superconducting tapes initiated by thermal disturbances

M Lebioda, J Rymaszewski[1]
Institute of Electrical Engineering Systems, Lodz University of Technology, Stefanowskiego 18/22, 90-924 Lodz, Poland

[1]E-mail: jacekrym@matel.p.lodz.pl

Abstract. The normal zone formation in superconducting elements during the current flow is a dangerous process, which may lead to the irreversible destruction of materials. Results of the experimental and simulation studies of the normal zone propagation process in 1G (BSCCO) and 2G (ReBCO) superconducting tapes are presented in this paper. The obtained measurement and simulation results confirm the key role of the metallic shield of HTSC tapes in the heat transfer and the thermal stability.

1. Introduction

A phenomenon of the normal zone formation in superconducting elements during the current flow is a dangerous process, which may lead to the irreversible destruction of materials (see figure 1). The evolution of the normal zone is caused by thermal, electrical or magnetic disturbances close to a superconductor. A local increase of resistance and, consequently, an increase of the generated Joule heat associated with the transport current flow are the results of this phenomenon [1]. This process is particularly dangerous in case of the high current devices.

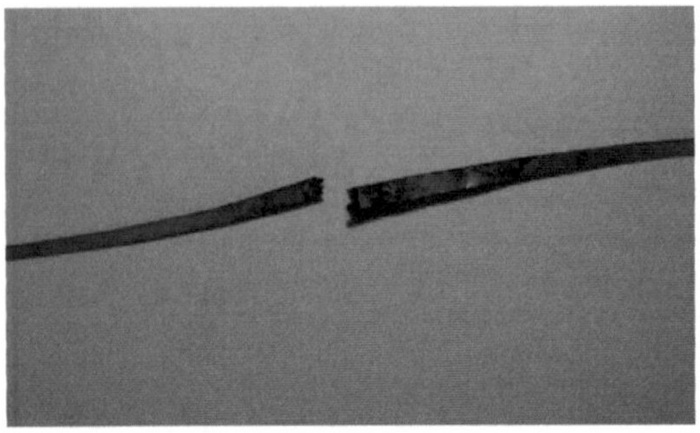

Figure 1. 2G superconducting tape burned out due to developed normal zone, initiated by thermal disturbance.

The operating current of these devices can be close to the critical current of superconducting tapes, especially during emergency situations. A metal stabilizer coating the superconductor is therefore very important in development of this zone [2,3]. A local destruction of superconductivity causes the transport current redirection from the superconductor to the metallic sheath. If the thermal conductance of the stabilizer is appropriately large the heat rapidly diffuses within the element. The high speed of the normal zone propagation velocity (NZPV) is therefore required to protect the superconducting materials from the destruction.

Designing of superconducting devices such as electromagnets or fault current limiters requires knowledge of the normal zone propagation in the windings [3-6]. Devices with superconducting coils, built so far, used the LTS superconducting wires, i.e. Nb_3Sn, Nb-Ti. The second generation (2G) superconducting tapes (ReBCO) actually become the most attractive material for the windings. 2G HTS tapes offer unprecedented mechanical (high tensile strength and mechanical stability) and electromagnetic parameters (very high critical current and upper magnetic field), especially in liquid nitrogen temperatures. The great advantage of the 2G tapes is high thermal stability resulting from the very small thickness of the superconducting layer in relation to a substantially thicker layer of metallic substrate. The temperature margin for coils made of high temperature superconducting materials is greater by an order of magnitude in comparison to the low temperature superconductors and the heat capacity is greater by 2-3 orders. Unfortunately, the NZPV in 2G tapes is on the order of 1 mm/s to 10 mm/s [2-5,7-9], which is very much lower than that of LTS wire (NZPV > 10 m/s) [10]. As a consequence, the heat dissipation doesn't relocate and the appearing hot spots could be very high and destructive in this case. A large resistance at the superconductor-stabilizer interface is a reason of the small zone velocity. All these features make it difficult to design the stable electromagnet with 2G coils.

The authors have plans to design and construct the superconducting electromagnet with HTS coils [11]. Therefore, the analysis of the normal zone propagation becomes necessary. Results of the experimental and simulation studies of the normal zone propagation process in superconducting tapes are presented in this paper. The 1G (BSCCO) and 2G (ReBCO) superconducting tapes, cooled in the liquid nitrogen, were examined and compared.

2. Experiment

A special sample holder, made of the nylon (polyamide) rod, is presented in figure 2. The tape with a length of 25 cm was wound around the rod. This design minimizes the influence of heat from the current electrodes to analyze the region of the tape. Many voltage taps were arranged on both sides of the samples' surface to examine the propagation process. The measuring system was prepared for measurements of the normal zone propagation velocity.

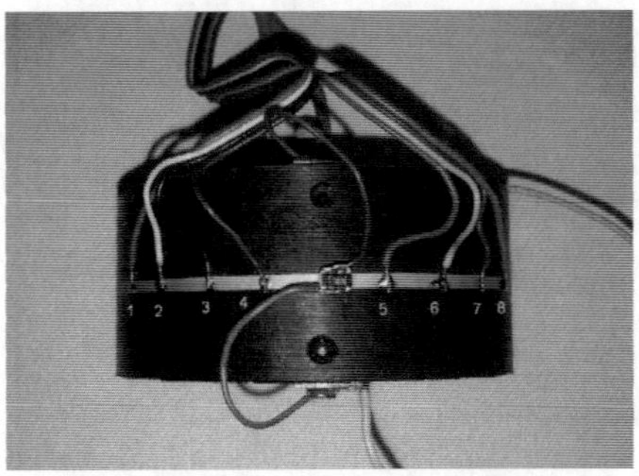

Figure 2. 2G sample tape with voltage taps and SMD resistor as a heater.

The main element of the system is the digital oscilloscope TDS5032B characterized by the high sampling speed (up to 5 GS/s) with the differential amplifier ADA400A, which enables the precise (with 1 μV resolution) acquisition of the signal on the voltage electrodes (probes) of the superconductor in the static and dynamic states. Both the transport current source and the heater current source are software controlled which allows full control over the process of the developing of the normal zone. Thermal disturbances were initiated by the heater made of SMD resistors 2512 (12.5 Ω). Two different types of superconducting tapes were examined: 1G – BSCCO 2223 – NST Zerome Hercules (width – 4 mm, thickness – 0.4 mm, measured critical current 50 A) and 2G –

ReBCO – SuperPower® SCS6050AP, sliced to a width approximately 2 mm (thickness – 0.1 mm, measured critical current 65 A). The sample holder with the tested tape was placed in a liquid nitrogen bath. The block diagram of the measurement system is presented in figure 3.

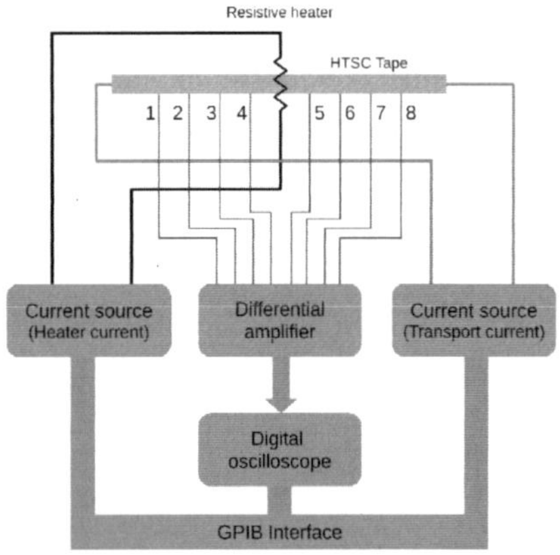

Figure 3. Block diagram of measurement system.

3. Results

Measurements were made for various parameters of the transport current and heating current sources (single or multi impulses). The following figures are merely examples of the obtained results. Figure 4 presents comparison of voltages' changes associated with the thermal local disturbance for both type of tapes: BSCCO and ReBCO. The local loss of stability for current lower than critical has been observed. It should be noted that the dynamics of changes and the voltage attained are higher in ReBCO sample. The thermal pulse applied on the samples is shown in the rapid increase of voltage. The rising of voltage is nearly linear and slight thermal fluctuations have been observed close to peak voltage. This could be the result of change of the boiling intensity. In case of the BSCCO sample a lower dynamic of voltage changes and non-linear rising of the voltage have been observed. This is the result of fast propagation of heat flux along the samples. The high thermal conductivity and the large thickness of BSCCO metallic stabilizer are the reasons for this behavior. The intense thermal fluctuations have not been observed in this case.

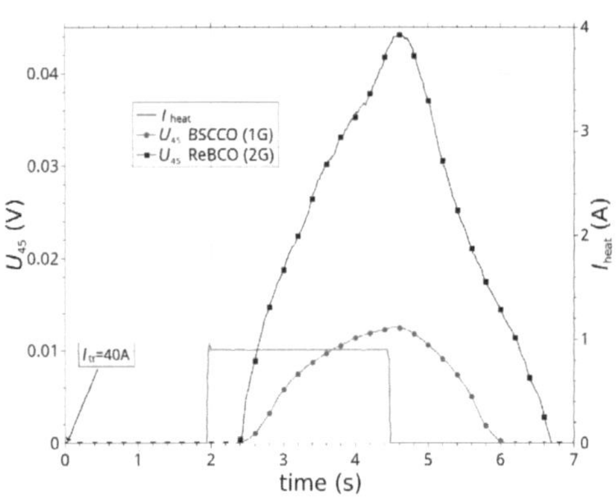

Figure 4. Voltage change on 1G and 2G samples caused by the thermal pulse (I_{heat} = 0.9 A, t_h = 2.5 s, E ~ 25 J). Transport current, I_{tr} = 40 A, lower than critical current.

Figure 5 shows time characteristics of the voltage between taps 4 and 5, localized near the heater on ReBCO tape. A much higher increase in voltage (over 100 mV on 20 mm) compared to the characteristics shown in figure 4 can be noted. This is caused by a larger amount of current transport (I_{tr} = 50 A) and higher pulse energy of heat (although shorter pulse duration). The voltage increase is

Journal of Physics: Conference Series **709** (2016) 012011 doi:10.1088/1742-6596/709/1/012011

due to the development of normal zone, which started near the heater. The thermal runaway was observed for the pulse duration t_h = 800 ms (energy $E \sim 35$ J). The positive feedback drives this process and this could end up burning out the tape if it does not comply with safety system (quench detector). Simultaneous measurements of the voltage between taps 4 and 5 as well as 5 and 6 showed that the resistive zone does not reach area between taps 5 and 6. The estimated value of the NZPV is relatively small and does not exceed 3 mm/s. For the shorter pulses this phenomenon does not occur and the voltage starts to decrease after reaching the maximum. It should be emphasized that a considerable part of the heat generated by the heater is absorbed by the liquid nitrogen.

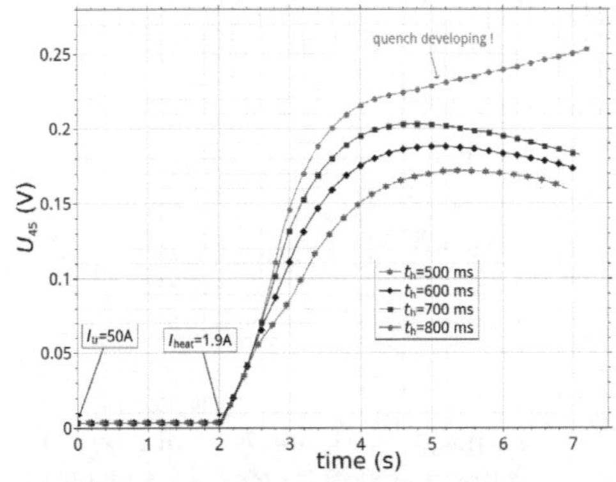

Figure 5. Time characteristics of voltage on 2G sample for thermal pulses with various width (I_{heat} = 1,9 A). Transport current, I_{tr} = 50 A, lower than critical current.

The low propagation velocity in 2G tapes favors the local accumulation of heat, which significantly affects the condition of heat exchange with the refrigerant. A significant shift in the time between current pulses in the heater and a voltage response of the sample is shown in figure 6. After every subsequent thermal pulse a marked increase in the voltage on the sample is observed. The low velocity of the resistive zone propagation leads to permanent loss of thermal stability.

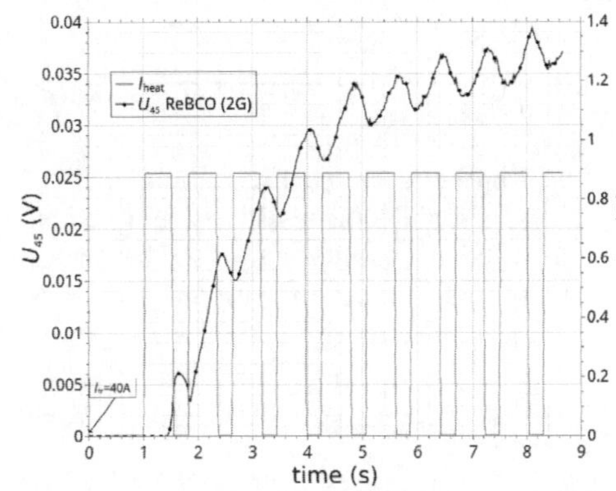

Figure 6. Voltage response on multiple thermal pulses for 2G sample (pulses width t_h = 0.5 s, transport current I_{tr} = 40 A).

4. Model

The model is implemented in COMSOL Multiphysics package. The main assumptions for the model are the following: electrical and thermal parameters of the materials are temperature dependent (including cryogenic temperatures), the boiling crisis has been included in the description (heat transfer coefficient is also temperature dependent) [11,12]. The equation of electrical conduction for solids and the equation of heat conduction are used. Other assumptions are described in the previous articles of authors [1,11-13]. The model allows, among others, to analyze local temperature changes of the superconducting tape, also just below the heater and in close proximity. Significant dimensions and the high thermal capacity of typical temperature sensors make proper measurements impossible. Figures 7 and 8 show typical simulation results. In figure 7, the process of heating of the tape, which is the effect of the heat pulse, can be observed. The effect of movement of the heating (resistive) zone (shifting of curves T_0 and T_1) is visible. The determined value of the NZPV is in this case approximately 50 mm/s. Despite the end of the heat pulse, the resistive zone continues to expand along the tape, which is caused by the heat generated by the transport current in the resistive area (figure 8). The direction of heat flow vectors confirms this assumption. The visible effect of the positive thermal feedback can lead to thermal damage of the tape (see figure 1).

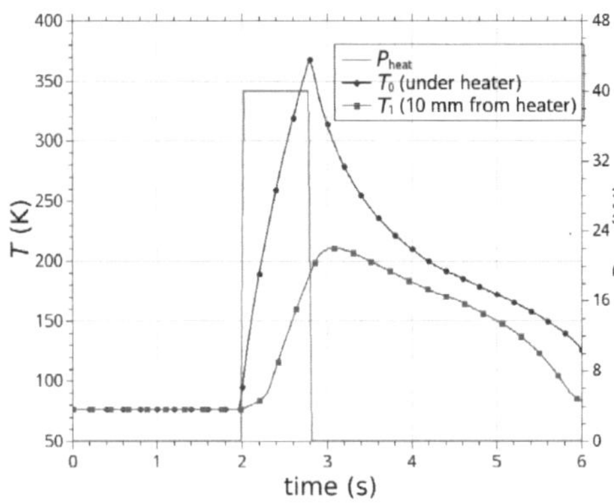

Figure 7. Time evolution of temperature of 1G tape as a result of heat pulse.

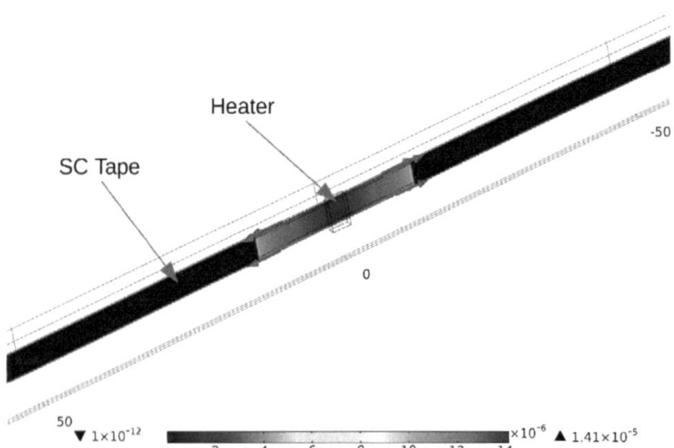

Figure 8. Distribution of superconductor resistivity and vectors of heat flux after end of the heat pulse ($t = 3$ s, transport current $I_{tr} = 80$ A, heat pulse energy $E_h = 32$ J).

5. Summary

This study investigated the development of the normal zone in two different types of superconducting tapes – 1G (BSCCO 2223) and 2G (ReBCO). The thermal local disturbance results in a change of electrical parameters of both types of HTSC tapes. A dynamics of local parameters' changes is high in case of 2G tape. This is the result of particular construction of the tape, where the thermal conductivity and capacity are slight. The development of the normal zone in this type of tape has been observed. The zones propagation after local heat source extinction has been noticed. The transport current maintains the heating process, which is the result of slow rate of the thermal resistive zone propagation. A local heat accumulation can be a main reason of damage of HTSC tapes. The coincidence between the experimental and simulation results and, in both cases, the occurrence of positive feedback between the transport current and the propagation of the normal zone have been observed. Experimental studies for 2G tapes have proven that the value of the NZPV is very low (<3 mm/s). The thermal runaway have not been observed for examined BSCCO tapes. The simulation research has shown that the NZPV for 1G tapes is about 50 mm/s. The proposed mathematical model allows the observation and analysis of the experimentally inaccessible phenomena. The obtained measurement and simulation results confirm the key role of metallic shield of HTSC tapes in heat exchanges.

6. References

[1] Rymaszewski J, Lebioda M and Korzeniewska E 2011 *Materials Science and Engineering* B: *Solid-State Materials for Advanced Technology* **176** 334-9
[2] Duckworth R C, Lue J W, Lee D F, Grabovickic R and Gouge M J 2003 *IEEE Transactions on Applied Superconductivity* **13** 1768-71
[3] Lacroix C, Fournier-Lupien J H, McMeekin K, and Sirois F 2013 *IEEE Transactions on Applied Superconductivity* **23** 4701605
[4] Colangelo D and Dutoit B 2015 *IEEE Transactions On Applied Superconductivity* **25** 5601708
[5] Roy F, Therasse M, Dutoit B, Sirois F, Antognazza L and Decroux M 2009 *IEEE Transactions On Applied Superconductivity* **19** 2496-2499
[6] Majka M, Kozak J, Kozak S, Wojtasiewicz G, Janowski T 2015 *IEEE Transactions on Applied Superconductivity* **25** 5601005
[7] Bae D K, Park D K, Ahn M C, Kang H, Yoon Y S and Ko T K 2006 *Journal of Physics: Conference Series* **43** 1043-4
[8] Zhong Z, Ruiz H S, Lai L, Huang Z, Wang W and Coombs T 2015 *IEEE Transactions on Applied Superconductivity* **25** 6602005
[9] Kim S B, Ueno Y, Ishiyama A, Okada H, Nomura S, Maeda H 1996 *IEEE Trans. Magn.* **32** 2822
[10] Wilson M N 1983 *Superconducting Magnets* (New York, NY, USA: Oxford Univ. Press) p 207
[11] Lebioda M, Rymaszewski J and Korzeniewska E 2014 *Journal of Physics: Conference Series* **494** 012018
[12] Lebioda M and Rymaszewski J 2013 *Przeglad Elektrotechniczny* **89** 280-3
[13] Lebioda M, Rymaszewski J and Korzeniewska E 2012 *Przeglad Elektrotechniczny* **88** 183-6

Charging/discharging processes in nanocrystaline MOS structures - Theoretical study

D Tanous, A Mazurak and B Majkusiak
Institute of Microelectronics and Optoelectronics,
Warsaw University of Technology, Koszykowa 75, Warsaw, Poland

E-mail: d.tanous@stud.elka.pw.edu.pl

Abstract. We present the study of impact of some parameters of the metal-insulator-semiconductor structure with nanocrystals embedded in the insulator layer on the current-voltage and capacitance-voltage characteristics with the bias voltage ramp rate as a parameter. The developed model is used as a tool for theoretical understanding the physics behind charging and discharging processes in the considered structures.

1. Introduction

Metal-insulator-semiconductor structures containing nanocrystals (nc) embedded in the insulator (figure 1a) have attracted scientific interest for their potential charge retention [1,2] and light emitting [3] properties. The two features constitute possible implementation fields: charging nodes in electron memory devices and integrated photonics. In this work C-V-t and I-V-t characteristics of such structures are analysed with the use of a theoretical model.

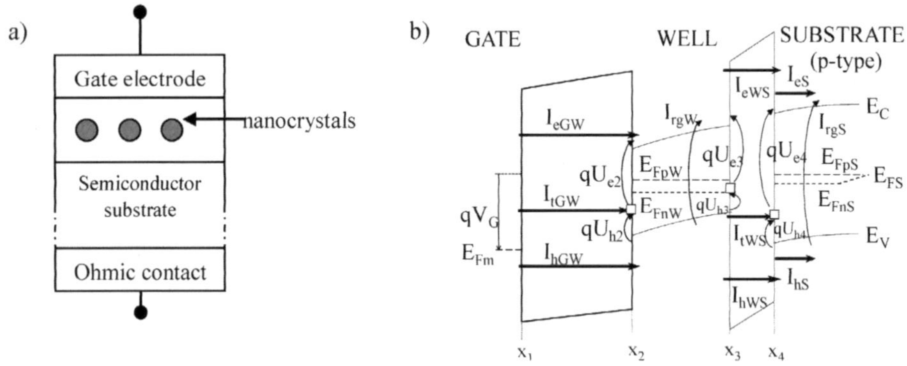

Figure 1. MIS structure with nanocrystals (ncMOS): cross-section (a), energy-band diagram (b).

2. Theoretical model

The energy-band diagram and current flow paths through the ncMOS structure with the p-type semiconductor substrate under the negative gate bias are shown in figure 1b. The steady state is satisfied by preservation of continuity of currents in the structure. When a steady-state is disturbed by a bias change, the electrostatics problem is solved under the assumption that charges accumulated in the well remain unchanged just after the bias jump. It means that the balance of fluxes is not fulfilled and the unbalanced fluxes ($\Delta J_{h/e}$) charge or discharge the well region with a time ratio $\Delta Q_{h/e}= t\cdot\Delta J_{h/e}$ until a new steady-state is acquired after sufficient number of Δt time incremental steps. After each time step Δt the electrostatics problem is solved for modified $Q_{h/e}$ values and the high-frequency capacitance $C = \mathrm{d}Q_G/\mathrm{d}V_G$ is calculated as the ratio of the response of the gate charge to the small incremental change of the bias voltage $\mathrm{d}V_G$ at the assumption that the electric charges in the well and the minority carrier charge in the substrate do not response to the $\mathrm{d}V_G$ change. The total current (I_{tot}) flowing through the structure can be expressed as:

$$I_{\mathrm{tot}} = I_{\mathrm{eS}} + I_{\mathrm{hS}} \tag{1}$$

where

$$I_{\mathrm{eS}} = I_{\mathrm{eWS}} + \mathrm{q}U_{\mathrm{e4}} = I_{\mathrm{eGW}} + \mathrm{q}U_{\mathrm{e2}} + I_{\mathrm{rgW}} + \mathrm{q}U_{\mathrm{e3}} + \mathrm{q}U_{\mathrm{e4}} \tag{2}$$

$$I_{\mathrm{hS}} = I_{\mathrm{hWS}} - \mathrm{q}U_{\mathrm{h4}} = I_{\mathrm{hGW}} - \mathrm{q}U_{\mathrm{h2}} - I_{\mathrm{rgW}} - \mathrm{q}U_{\mathrm{h3}} - I_{\mathrm{tWS}} - \mathrm{q}U_{\mathrm{h4}} \tag{3}$$

I_{eS}/I_{hS} are the electron/hole currents at the substrate surface, $\mathrm{q}U_e/\mathrm{q}U_h$ are the electron/hole currents between the interface traps and the bands, I_{eGW}/I_{hGW} are the electron/hole tunnel currents between the gate and the well, I_{eWS}/I_{hWS} are the electron/hole tunnel currents between the well and the substrate, I_{tWS} is the tunnel current between the substrate interface traps and the nanocrystal bands, I_{rgW} is the thermal recombination-generation current in the well region. Densities of the electron/hole tunnel currents are calculated according to the Tsu-Esaki model [4]:

$$J_{\mathrm{e/h}} = \mathrm{q} \int_{\mathrm{C/V\,band}} P(E_x) N(E_x) \mathrm{d}E_x \tag{4}$$

where the integration is over allowed states at both sides of the insulator. If the energy level of the carrier corresponds to the energy gap at any side of the insulator, it is excluded from the integration. For example, the "supply function" $N(E_x)$ for electrons tunneling through the substrate insulator (creating the current J_{eWS}) can be expressed by the formula [4,5]:

$$N(E_x) = \frac{4\pi \mathrm{m}_{2D} \mathrm{k}T}{\mathrm{h}^3} \ln\left(\frac{1+\exp[(E_{\mathrm{FnS}} - E_x)/\mathrm{k}T]}{1+\exp[(E_{\mathrm{FnW}} - E_x)/\mathrm{k}T]} \right) \tag{5}$$

where m_{2D} is the two-dimensional density-of-states effective mass and the tunneling probability $P(E_x)$ is equal to [5]:

$$P(E_x) = P_0 \exp\left(-2 \int_{x_3}^{x_4} \kappa(x) \mathrm{d}x \right) \tag{6}$$

where the pre-exponential factor P_0 has the form:

$$P_0 = 16 \frac{\frac{k_{W3}}{m_{xW}} \frac{\kappa_{i3}}{m_{ie}} \frac{\kappa_{i4}}{m_{ie}} \frac{k_{S4}}{m_{xS}}}{\left[\left(\frac{k_{W3}}{m_{xW}}\right)^2 + \left(\frac{\kappa_{i3}}{m_{ie}}\right)^2\right]\left[\left(\frac{\kappa_{i4}}{m_{ie}}\right)^2 + \left(\frac{k_{S4}}{m_{xS}}\right)^2\right]} \tag{7}$$

m_{xW} and m_{xS} are the longitudinal effective masses in the well and the substrate, respectively, m_{ie} is the electron effective mass in the insulator, k_{W3} and k_{S4} are the wave vectors at the right side of the well and at the semiconductor substrate, respectively, while κ_{i3} and κ_{i4} are the imaginary wave vectors in the insulator at the turning points x_3 and x_4, respectively. The imaginary wave vector $\kappa_i(x)$ is determined at the zero value of the transverse component $E_\perp$ according to the one band barrier model and is given by:

$$\kappa_{i3,4} = \left[\frac{2m_i}{\hbar^2}(E_{Ci3,4} - E_x)\right]^{1/2}, \tag{8}$$

The tunneling currents J_{eGW}, J_{hGW} and J_{hWS} are computed analogously. Rates of charge exchange between the interface traps and semiconductor conduction U_c and valence U_v bands at the specific surface are given by [6]:

$$U_c = q\sum_i N_{it}(E_{it})\upsilon_{thn}\sigma_{tn}[n_s - f_t(n_s + n_1)] \tag{9}$$

$$U_v = q\sum_i N_{it}(E_{it})\upsilon_{thp}\sigma_{tp}[f_t(p_s + p_1) - p_1] \tag{10}$$

where σ_{tn} and σ_{tp} are capture cross sections for electrons and holes, υ_{thn} and υ_{thp} are the average thermal velocities of electrons and holes, respectively, p_s and n_s are the electron and hole concentrations at the interface, and p_1 and n_1 are the characteristic trap parameters. The thermal generation-recombination currents J_{rgW}, J_{eS} and J_{hS} are calculated by integration of the SRH [7] net recombination rate over the well and substrate regions. For example:

$$J_{rgW} = q\int_{x_2}^{x_3} r_{thW}(x)dx = kT\int_{u_3}^{u_2} \frac{r_{thW}(u, u_{FpW}, u_{FnW})}{F(u, u_{FpW}, u_{FnW})}du \tag{11}$$

where

$$r_{thW}(u, u_{FpW}, u_{FnW}) = \frac{n_i}{2\tau_0}[\exp(u_{FpW} + u_{FnW}) - 1]\cdot\left[1 + \frac{\tau_{p0}e^{u+u_{FnW}-u_F} + \tau_{n0}e^{-u+u_{FpW}+u_F}}{2\tau_0}\right]^{-1} \tag{12}$$

with the normalized potential u=qΦ/kT and normalized splits of the quasi-Fermi level from the equilibrium level in the corresponding region.

3. Discussion

The material parameters considered in the simulations are marked in figure 2 and the bias voltage ramp scheme is presented in figure 3. A nanocrystal is represented in the model by a potential well. Several values of the energy gap E_{gW} and the electron affinity E_{CW} are considered in simulations in order to include the energy quantization effect in the well. Therefore, solution to Schrödinger's equation is avoided in the calculations what improves computation efficiency.

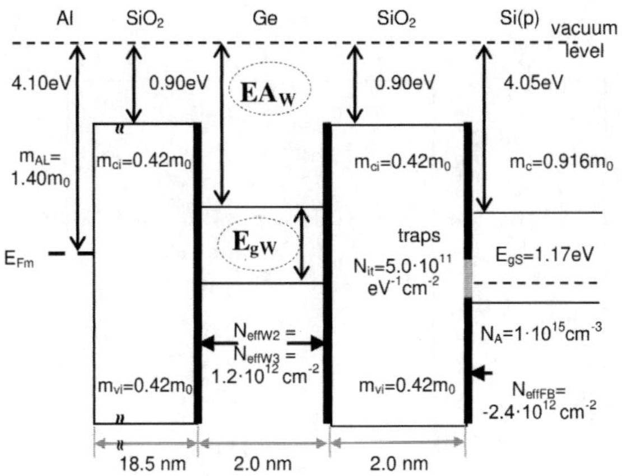

Figure 2. Parameters of the ncMOS structure assumed in simulations.

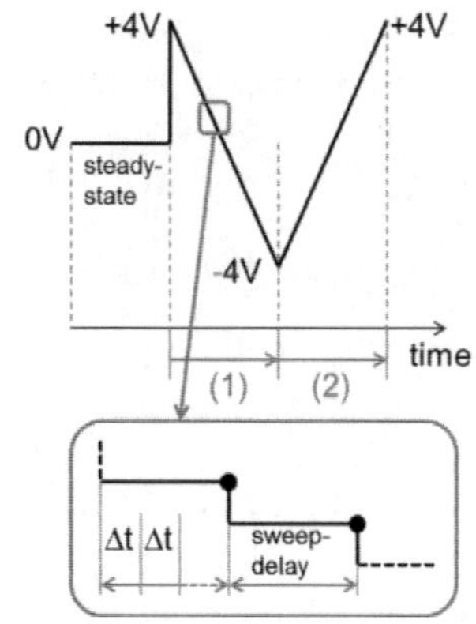

Figure 3. The bias voltage ramp.

The steady-state means that equilibrium electrostatics is acquired i.e. in regard to the structure in figure 2 there is no fluxes flow in the structure. The effects of the electron affinity (E_{CW}) and band gap (E_{gW}) of the Ge nanocrystals (ncGe) on the memory window width are shown in figure 4, 5. At the positive biases, the potential well in the nanocrystal region is charged by electrons tunneling from the substrate (current I_{eWS}) and the CV curve rising arm is shifted to the right (figure 4a) proportionally to the stored negative charge in the well. At the negative biases, the nanocrystal region is charged positively by holes tunneling from the substrate (current I_{hWS}) and the CV curve is shifted to the left. The falling arms of three CV curves in figure 4a for the reversed bias sweep overlap since they correspond to the same value of the level E_{VW} determining the bottom of the potential well for holes. The current in this bias range is determined mainly by holes (current I_{hWS}) charging or discharging the well (figure 4b).

Changes of the level E_{VW}, which can result from changes of the nanocrystal energy gap E_{gW} for the established value of the electron affinity E_{CW} (see figure 5), affect mainly the process of charging and discharging the nanocrystal by tunnel current of holes.

The simulation results can qualitatively comment the experimental data which can be found in references [8,9].

a) b)

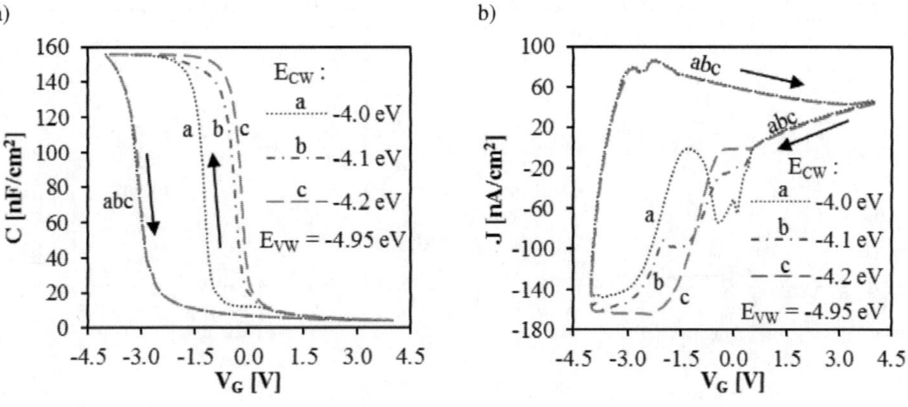

Figure 4. C-V-t (a) and I-V-t (b) characteristics for the ncGe-MOS: E_{CW} as a parameter (E_{gW} = 0.67 eV).

Journal of Physics: Conference Series **709** (2016) 012012 doi:10.1088/1742-6596/709/1/012012

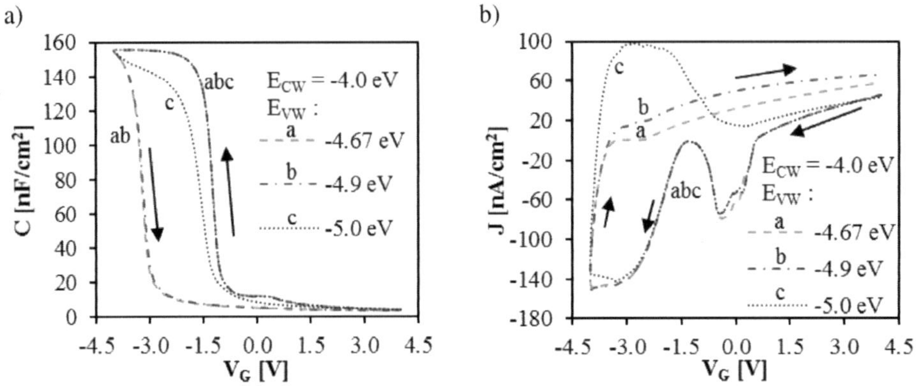

Figure 5. C-V-t (a) and I-V-t (b) characteristics for the ncGe-MOS: *EG* as a parameter (EA = 4.0 eV).

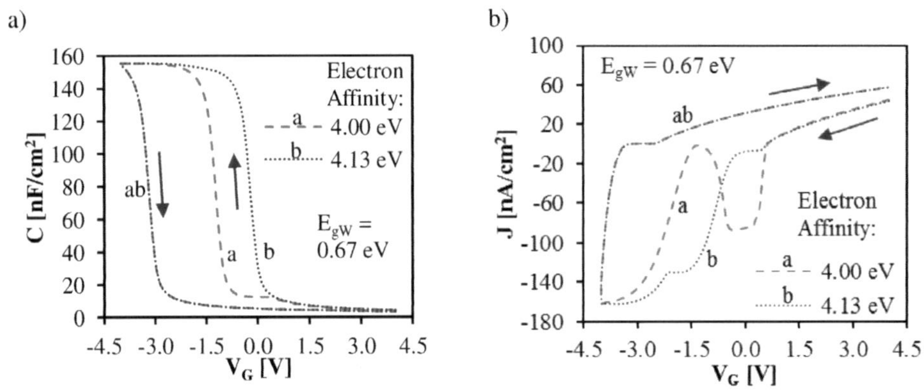

Figure 6. C-V-t (a) and I-V-t (b) characteristics of the ncGe-MOS: *EA* as a parameter (E_{gW} = 0.67 eV).

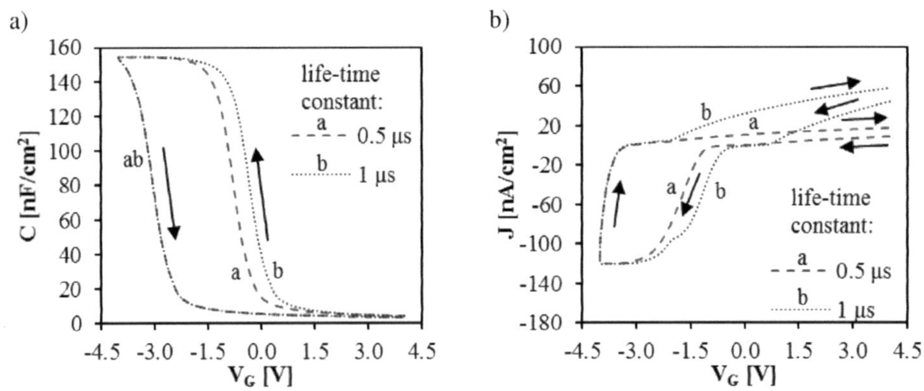

Figure 7. C-V-t (a) and I-V-t (b) characteristics of the ncGe-MOS: substrate carrier life-time constant as a parameter.

Journal of Physics: Conference Series **709** (2016) 012012 doi:10.1088/1742-6596/709/1/012012

A literature review proves that the ncGe physical parameters, in respect to the bulk ones, are open to doubt. The differences between electrical characteristics (C-V-t and I-V-t) for commonly proposed in literature Ge electron affinity values (i.e. 4.00 eV [10], 4.13 eV [11]) are shown in figure 6. The ncGe band gap E_G is also a subject to discussion, e.g. [8]. For the simulations mentioned above the Ge bulk value (0.67 eV) was assumed. The effect of the electron lifetime in the substrate is presented in figure 7. The lower lifetime results in a higher generation rate, which in turn enhances significantly the electron charging of the well (as it is limited by the minority carrier generation). The memory window width is also affected by the bias sweep rate as presented in figure 8. At each bias point the structure tends to reach the steady-state (solid line in figure 8a).

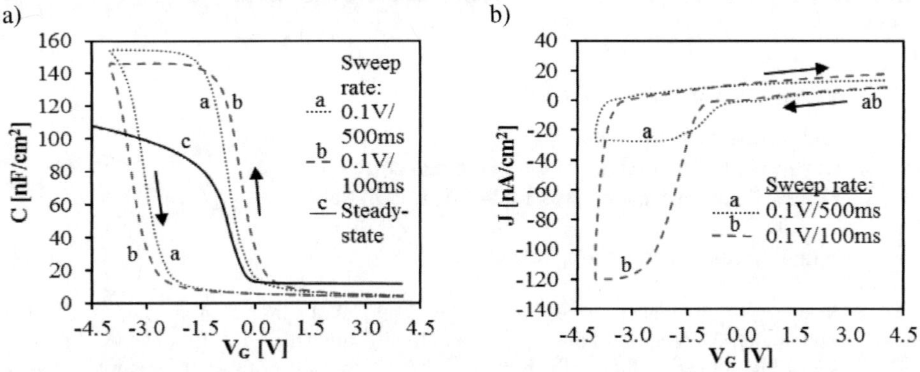

Figure 8. C-V-t (a) and I-V-t (b) characteristics for two different bias voltage ramp rates.

4. Conclusions

In this paper, the time-dependent capacitance and current responses were investigated for the ncMOS structures with nanocrystals (Ge) embedded in SiO_2. The considered parameters have strong impact on dominant current paths in the selected bias regions. The conducted study and the developed model may be useful for electrical characterization and fabrication process optimization. Fitting the simulated C-V-t and I-V-t characteristics to experimental results can prove the physical structure, parameters and the position of the nanocrystals in the insulator.

Acknowledgments
This work was financed by National Science Centre funds dedicated upon the decision no. DEC 2011/03/B/ST7/02595.

References

[1] Bez R and Pirovano A 2004 *Mat. Sci. Semicond. Process.* **7** 349–55
[2] Mroczynski R *et al* 2012 *Appl. Surf. Sci.* **258** 8366–70
[3] Yer S S and Xie Y H 1993 *Science* **260** 40
[4] Tsu R and Esaki L 1973 *Appl. Phys. Lett.* **22** 562
[5] Majkusiak B and Jakubowski A 1985 *J. Appl. Phys.* **58** 3141-4
[6] Majkusiak B 1998 *IEEE Trans. Electron Devices* **45** 1903-11
[7] Shockley W and Read W T 1952 *Phys. Rev.* **87** 835-42
[8] Beyer V *et al* 2008 *J. Appl. Phys.* **106** 064505
[9] Dominik T, Mazurak A and Majkusiak B 2014 *Theoretical and Experimental Investigation of ncMOS Structures with Ge nanocrystals for Memory and Photonics Applications* 18th Workshop on Dielectrics in Microelectronics 9-11 June 2014 Kinsale
[10] Sze S M and Ng K K 2006 *Physics of Semiconductor Devices* Third Edition (New Jersey: Wiley–Interscience)
[11] Robertson J 2013 *J. Vac. Sci. Technol.* A **31** 050821

Influence of the quantum well models on the numerical simulation of planar InGaN /GaN LED results

J Podgórski, J Woźny, Z Lisik

Lodz University of Technology,
Department of Semiconductor and Optoelectronic Devices
211/215 Wolczanska Street, 90-924 Lodz, POLAND

E-mail: jacek.podgorski@p.lodz.pl

Abstract. Within this paper, we present electric model of a light emitting diode (LED) made of gallium nitride (GaN) followed by examples of simulation results obtained by means of Sentaurus software, which is the part of the TCAD package. The aim of this work is to answer the question of whether physical models of quantum wells used in commercial software are suitable for a correct analysis of the lateral LEDs made of GaN.

1. Introduction

Numerical simulations are the tool that is often used to investigate and optimise new semiconductor devices. This approach is especially useful when the semiconductor devices that must be characterised by the special dimensions and sophisticated design. The CAD packages basing on physical description of phenomena taking place in semiconductor structures are an effective tool both at the stage of new devices design and optimisation of their parameters. There are a lot of homemade software packages that have been reported in literature, but nowadays, the application of the commercial, more universal ones is preferred. Among such packages, present on the market, one should mention Atlas-Silvaco, tiberCAD-TiberLab, Medici-Synopsys, GTS-GlobalTCADSolutions and Sentaurus. The latter one has been chosen as a numerical tool for the investigation presented in the paper.

Sentaurus package, initially, has been designed as a CAD tool for silicon technology. Next, following the progress in microelectronics, it was adopted for other materials and updated by models introducing new phenomena e.g. quantum effects. The modifications enlarged the extent of the software applicability in the formal way, but the packages were still used to investigate silicon structures mainly with limited credibility to other semiconductor materials. Therefore, in order to recognise the usefulness and correctness of Sentaurus code with implemented models of physical phenomena for modelling of gallium nitride (GaN) structures, series of test simulations were elaborated. They focused on a simple real optoelectronic GaN structure. Special attention has been paid to modelling the quantum wells (QWs) appearing in the considered domains. In the Sentaurus code, QWs can be treated without any quantum corrections, like in [1], or with the aid of quantum well models available in the Sentaurus libraries [2].

As the representative real optoelectronics GaN device for the numerical investigations, the planar InGaN/GaN blue LED structure covering MQW layer, which design and manufacturing is described in [3], has been chosen. The paper [3] presents most of the LED design data necessary for numerical model creation and the basic characteristics of manufactured device. In the investigations reported here, the numerical models of the chosen LED structure taking into account different QW approaches have been compared. The conducted simulations allow evaluating the available in Sentaurus QW models.

2. Numerical Models of QW

In the Sentaurus package, there are implemented two different physical models of quantum wells. The first *QWTransport* is based on a specific approach to the area of quantum well (QW). This area is considered as a point source of recombination [2, 4]. This assumption leads to a simplification of the phenomena occurring inside the QW, and forces the creation of an appropriate discretization grid, as shown in figure 1.

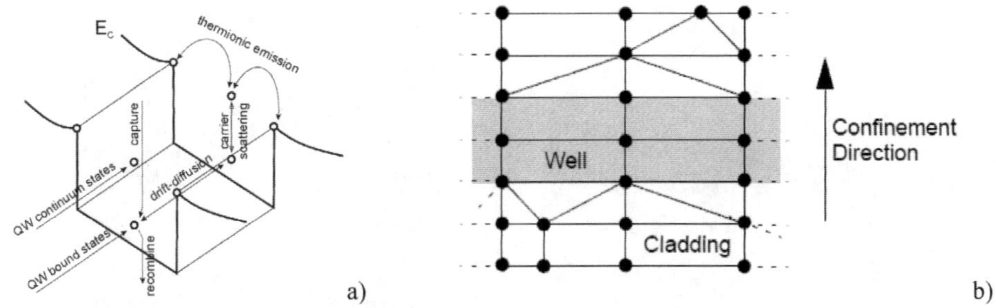

Figure 1. Discretization of quantum well according to Sentaurus *QWTransport* rules (physical phenomena – a, discretisation mesh – b)

Let us consider two cases of carrier transport by the quantum wells. In the case of transport perpendicular to the QW plane, the QW is preserved as a point source of recombination, which contains separate continuum and bound states. In this case, the continuum and bound states are assumed to have different quasi-Fermi levels that lead to separate continuity equations for the continuum and bound states. Transport of carriers from the regions outside the QW to the QW continuum states is by thermionic emission. The transition from the QW continuum to the bound states is computed by a scattering rate that includes carrier–carrier scattering. The bound states are solved from the Schrödinger equation. In the case of transport in a direction parallel to the plane of the QW, it is simulated basing on drift-diffusion model.

This approach determines a special mesh generation. The grid is based on the 'three points' model. It specifies that only one node of the mesh is placed in the centre of the well while the other two, respectively, at the edges of the QW region. This principle is schematically shown in figure 1b. Disadvantage of this approach is that there is no possibility to analyse the structures in which the QW regions are not perpendicular to the direction of carriers transport or when the QW region has more than one plane. The solution to the above mention problem is the application of another model – *QWLocal* [2]. From a technical point of view, this model does not require exactly 3 nodes in the quantum well. It is also possible to model structures in which the quantum wells can be arranged in different planes. This model has only one condition. The area of quantum well should be the same throughout the semiconductor structure.

In opposite to the *QWTransport* model in the *QWLocal* model, there are not designated carrier density corresponding to the status related to the quantum well. The carrier concentrations in the quantum well, as in the *QWTransport* model, are based on one-dimensional Schroedinger equation of energy levels - thereby modifying classically calculated carrier concentrations. The amount taking into account energy levels to analyse occurrence, is equal to: the sub-bands for electrons: 5, sub-band for light holes: 4, sub-bands for heavy holes: 2, sub-bands for the hole resulting from cleavage originating from the crystal field: 2, the amount of valence band: 3, respectively. In the case of LED modelling, there is no stimulated emission and spontaneous emission that takes into account the presence of the aforementioned sub-bands.

3. Modelling of Real GaN LED Structure

The models presented above confirmed that in general, we are able to use the Sentaurus package as the base to model phenomena taking place in optoelectronics GaN structures. We decided to perform simulations of a real GaN blue LED described in the literature [3]. The chosen structure is shown in photographs [3] in figure 2 presenting several chips on one wafer and the blue light emitted from one of the fabricated chips, respectively.

a) b)

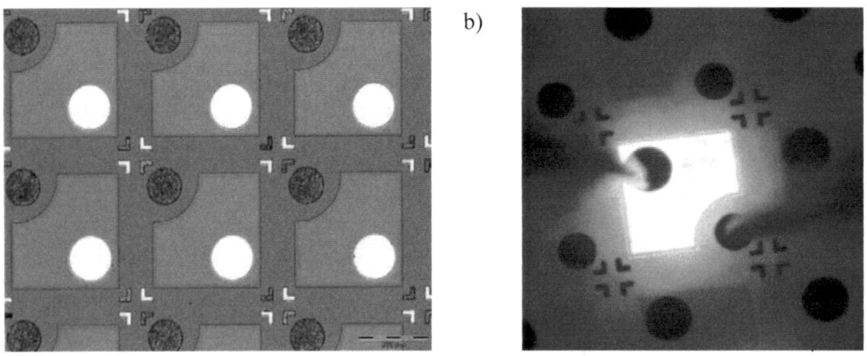

Figure 2. Photographs of several LED chips (a) and blue light emitted from one of the fabricated LED chips (b) [3]

It is a planar rectangular structure with dimensions 315 x 315 μm^2 and two round contacts placed along the diagonal at opposite corners of the structure, as it is sketched by the white line in figure 3a. The diagonal represents the simplest way for the current flow from anode to cathode, characterised by the lowest electrical resistance. It means that in the plane defined by the white axis, the direction of the anode-cathode current exactly agrees with the direction of the white axis and its flow can be considered as 2D phenomenon. It has allowed to develop a 2D model for its investigation, in which, the numerical simulations are carried out for the 2D domain corresponding to plane defined by the white axis. The layers that have been distinguished in the 2D domain together with their lateral dimensions are shown in figure 3b.

The sample LED consists of several layers. These include: sapphire substrate, GaN buffer layer, Si-doped n-GaN layer, multi quantum wells layer, cap GaN layer and Mg-doped p-GaN layer. The design data necessary to develop the numerical model of considered structure has been taken from the study [1] mainly and completed on the base of literature research. They are collected in table 1. Not all the layers listed in the table have finally been taken into account in the developed model. It covers the layers 3-7 only. The contact layers 1-2 and 10 have been replaced by ideal contacts whereas the layers 8-9 have been removed due to their negligible participation in the anode-cathode current flow.

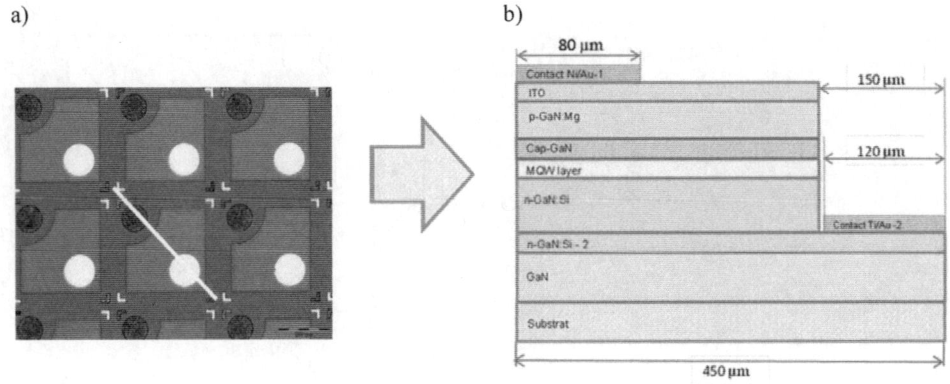

Figure 3. LED structure: (a) and the top view of the diagonal location on the considered chip [3], and (b) 2D domain corresponding to the diagonal plane

Table 1. Data describing the considered real GaN blue LED structure

	Layer	doping [cm^{-3}]	thickness [nm]
1	Contact-Ni/Au1	-	200
2	ITO	-	100
3	p-GaN:Mg	$8 \cdot 10^{17}$	250
4	Cap-GaN	$1 \cdot 10^{16}$	30
5	MQW 3 pairs $In_{0,4}Ga_{0,6}N$ GaN barrier	$1 \cdot 10^{16}$	39 3 10
6	n-GaN:Si	$8 \cdot 10^{18}$	700
7	n-GaN:Si-2	$8 \cdot 10^{18}$	300
8	GaN		
9	Substrat		
10	Contact Ti/Au2		

4. Results

The worked out model has been used for isothermal (T=300K) simulations that run successfully. Simulations were conducted using both previously described models of quantum wells. For comparison, performed simulations also took into account the models of classical physics. The main aim of these simulations was to determine the basic parameters characterising the device, such as threshold voltage, the resistance in conducting state and the wavelength of the emitted radiation. The first two parameters were determined from the current-voltage (IV) characteristic, while the last from the emission characteristic. Both these characteristics are shown in figure 4a and figure 5a, respectively.

The first simulations were performed for the *QWTransport* model. Assuming the composition of $In_{0,4}Ga_{0,6}N$, compatible with the literature data [3] we obtained the maximum luminous intensity for wavelength λ = 545 nm and not as expected for 450 nm. Therefore, new fraction of the InGaN was calculated to fit the expected wavelength, assuming that the radiative recombination occurs only in quantum wells. For the final simulations we used the new ratio of indium fraction, which was 0.2. The same factor was used for subsequent simulations with other physical models (*QWLocal* and *NoQW*).

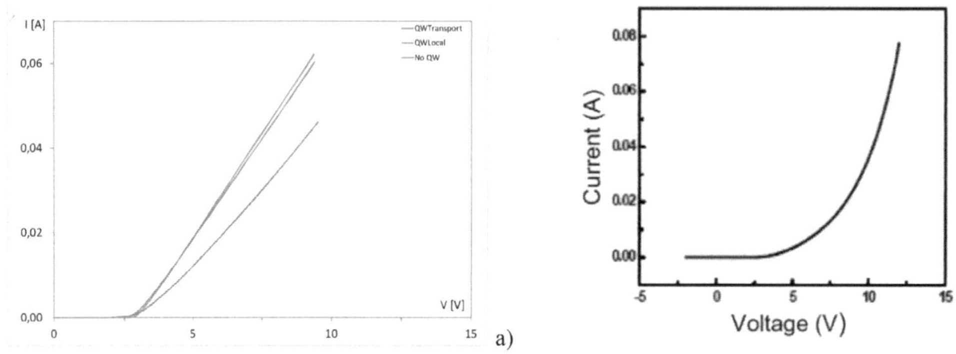

Figure 4. Comparison of simulated (a) and measured in [3] (b) I-V characteristics of considered GaN LED structure

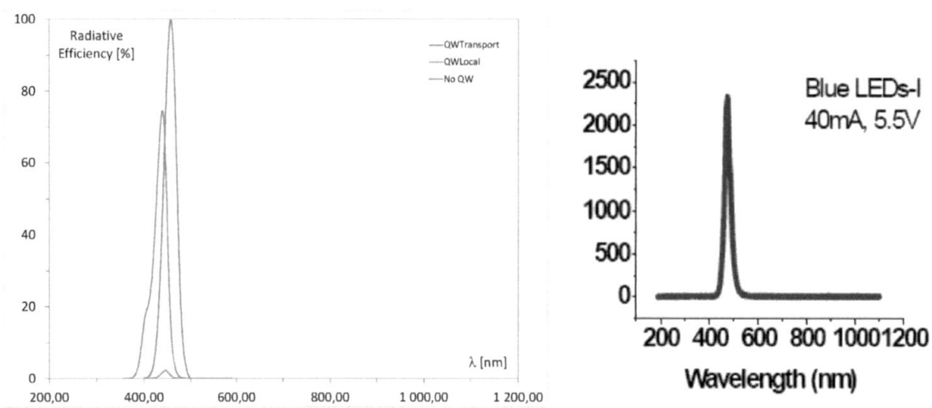

Figure 5. Comparison of the electroluminescence spectra resulted from numerical simulations (a) and measured in [3] (b) of considered GaN LED structure

Some of the results that have their counterparts in the measurement results presented in [3] are compared in figure 4 and figure 5. In the first case, the I-V characteristics, calculated as isothermal ones, are juxtaposed with the characteristic measured in non-isothermal conditions whereas in the second case the electroluminescence spectra are compared. Moreover, three extracted parameters from the above characteristics for three approaches are gathered in table 2. The threshold voltages V_T for numerical simulations are determined based on the linear approximation of I-V characteristics, while the wavelengths λ are defined as the maximum of the radiative efficiency spectrum. The measured values are taken from [3].

The similarity of numerical and experimental results is evident, e.g. the wavelength corresponding to the maximum of electroluminescence obtained numerically and experimentally are 440 nm or 445 nm and 466,5 nm, respectively. Full width at half maximum (FWHM) parameter, that allows determining the selectivity of the emitted signal in whole spectrum, is similar for *QWLocal*, *NoQW* approaches and the measured characteristic. Unfortunately, *QWTransport* suffers from unexplained emission suppression. Furthermore, the threshold voltage values obtained numerically and experimentally are comparable. The difference in the shape of the I-V curve can be caused by the fact that the numerical simulations were carried out under isothermal conditions (T = 300 K). It was not possible to determine the real measurement thermal conditions from [3]. The second factor that can influence the IV characteristic shape is the construction of the numerical structures and the real LED. The models do not include ITO current spreading layers. Removal of this layer affects the current distribution inside the structure, and in consequence current densities on contacts. In figure 5a, we can see that the *QWLocal* peak broadens. This is due to the occurrence of radiative recombination in the GaN structure on p-n junction, outside the multi-quantum wells region. Nevertheless, both types of presented characteristics show that the two tested quantum models behave similarly. The emission suppression in the case of *QWTransport* approach needs further investigations.

Table 2. Characteristic parameters InGaN LED obtained from simulation and measurements

	QWTransport	QWLocal	NoQW	measured
V_T [V]	3,2	3,1	3,5	4 [3]
λ [nm]	445	440	458	466,5 [3]
FWHM [nm]	20,7	29	31	~30

5. Conclusions

The paper focuses on verification of quantum models embedded in Sentaurus TCAD package. The main aim was to point out the most suitable approach to analyse optoelectronic GaN devices with quantum wells. The two models were tested. The results obtained using these models are comparable with the results obtained for the real LED [3]. Although both models behaved similarly while performing electrical analysis, the optical analysis of *QWTransport* model suffers from unexplained emission suppression. Moreover, *QWTransport* model has a very big disadvantage connected with specific requirements on grid generation. Hence, the *QWLocal* model should in the area of interest in the case of optoelectronic GaN device analysis.

In the case of commercial software, like Sentaurus, the user should pay attention the data delivered with built-in material libraries. The parameters defined in them may differ from the values given in other sources a lot. It may reflect the obtained results, so one should not treat them uncritically.

6. References

[1] Saguatti D, Bidinelli L, Verzellesi G, Meneghini M, Meneghesso G, Zanoni E, Butendeich R, Hahn B, 2012 *IEEE Trans. on Electron Devices* **59(5)** 1402-1409

[2] Sentaurus Device, User Manual, Ver. J_2014.09

[3] Nguyen X L, Nguyen T N N, Chau V T and Dang M C 2010 *Adv. Nat. Sci.: Nanosci. Nanotechnol.* **1** 025015 (5pp)

[4] Grupen M, Hess K 1998 *IEEE Journal of Quantum Electronics* **34(1)** 120-140

Acknowledgements

This work has been financially supported by the EU projects "GECCO" (Project No. 280694).